Bruno Flório Lessi
Fernando A. Bataghin
José S. R. Pires

Urban afforestation on the UFSCar campus, São Carlos - SP

Bruno Flório Lessi
Fernando A. Bataghin
José S. R. Pires

Urban afforestation on the UFSCar campus, São Carlos - SP

Management

ScienciaScripts

Cover image: www.ingimage.com

This book is a translation from the original published under ISBN 978-3-330-77097-3.

Publisher:
Sciencia Scripts
is a trademark of
Dodo Books Indian Ocean Ltd. and OmniScriptum S.R.L publishing group

120 High Road, East Finchley, London, N2 9ED, United Kingdom
Str. Armeneasca 28/1, office 1, Chisinau MD-2012, Republic of Moldova, Europe
Managing Directors: Ieva Konstantinova, Victoria Ursu
info@omniscriptum.com

Printed at: see last page
ISBN: 978-620-8-63436-0

SUMMARY

To my parents and family My partner Grabriela Friends and colleagues Dedication

PREFACE

We all know how nice it is to walk along streets and parks shaded by beautiful trees: the air is fresher, the temperature is milder, there is more protection from the sun, birds are around, fruits and flowers give us colors and smells. Among all the elements of vegetation found in a city, the set of trees present on sidewalks, streets, parks and squares is called urban afforestation. This element plays an important role in providing a wide range of economic, ecological and social benefits.

This publication provides an analysis of the urban tree planting on the *campus of* the Federal University of Sao Carlos, Sao Carlos - SP, which began by carrying out a georeferenced tree inventory to help with diagnosis, planning and management. The tree survey included information such as species names, origin, conflicts, physical damage, size, planting, plant health, integrity and functionality. But this publication goes far beyond numbers, providing an ecological analysis of the composition of urban tree species, as well as a technical analysis of tree quality, conflicts and management. At the end of the study, we provide some guidelines for planning in order to improve this arborization and maximize its benefits to *campus* users.

We would also like to remind you that the discussions and methodologies covered in this publication can be implemented at all scales, from a simple park to a city, or a group of cities, always remembering that the vegetation found in the city is not isolated, it is part of the urban ecosystem and is interconnected with the entire surrounding ecosystem and for this reason must be implemented and managed with criteria, taking into account the benefits for people and the ecology of the place.

CHAPTER 1

1 GENERAL INTRODUCTION

Currently, approximately 53% of the world's population lives in urban areas (PRB, 2014). Urban green spaces make an important contribution to residents' quality of life (BOLUND; HUNHAMMAR, 1999; GÓMEZ-BAGGETHUN et al., 2013; KONIJNENDIJK et al., 2005; NICHOLSON- LORD, 2005). In this context, quality of life is defined by Schwab (2009) as "the sum of all the things that make life enjoyable and meaningful, including physical, mental, economic, psychological, aesthetic and leisure benefits". In times of environmental crisis, especially with reference to global warming, good quality green infrastructure becomes even more important, given that vegetation mitigates the effects of climate change in urban areas and improves the quality of life for people who live and work in these areas (BOLUND; HUNHAMMAR, 1999; GILL et al., 2007).

The urban environment has been the target of many ecological studies aimed at understanding its functioning and ecological processes in order to improve it, making it more balanced, resilient and less aggressive to nature (MCPHERSON; NOWAK; ROWNTREE, 1994; ALBERTI, 2008; ELMQVIST et al. 2013). Urban ecology treats cities as an ecosystem (MCPHERSON; NOWAK; ROWNTREE, 1994; ALBERTI, 2008) and, as such, urban vegetation, even if it is often planted, or a combination of planted and native vegetation left over from urbanization processes, is considered extremely important for the ecological balance of this ecosystem. Urban vegetation has its herbaceous, shrubby and arboreal extract, where the shrubby and arboreal extract in the middle of the city, streets and green areas can be considered as the city's Urban Afforestation (MILANO, 1988; LIMA et al., 1994; MAGALHÂES, 2006).

With urban planning processes tending to remove all natural vegetation from the environment, the destination for these areas ends up being limited, in the midst of urban infrastructure. Due to the fact that this vegetation is in the middle of the urban environment, an artificial environment built by man according to his interests, we can observe the emergence of conflicts with the infrastructure of this environment

(MASCARÓ; MASCARÓ, 2010). Conflicts can arise between urban vegetation and public lighting, where a tree can hinder lighting with the growth of its crown. Conflicts can arise with constructions and buildings, where a tree may be damaging them because it is too close; conflicts can arise with the electricity grid, and with many other infrastructures existing in the urban environment. These conflicts can be caused both by the lack of infrastructure for tree planting and by the wrong choice of plant species (MASCARÓ; MASCARÓ, 2010). The presence of conflicts generates the need for tree management, which, if carried out incorrectly, can lead to permanent damage and defects in the trees (MASCARÓ; MASCARÓ, 2010).

Among the different elements of urban vegetation, trees are very important, playing a greater role than other plants in providing a wide range of economic, ecological and social benefits (KONIJNENDIJK et al., 2005). A number of studies show the importance of urban tree planting as a provider of these benefits (BOLUND; HUNHAMMAR, 1999; HIEMSTRA; BIJL; TONNEIJCK, 2008; NUFU, 2005), and show that these effects are multiplied by mature trees with a large canopy (BRITT; JOHNSTON, 2008).

The materials used in building construction store heat, raising the temperature of cities compared to natural areas (BIAS; BAPTISTA; LOMBARDO, 2003). Urban afforestation can influence climate improvement, contributing to human comfort (BOLUND; HUNHAMMAR, 1999; CARNEIRO et al., 2007).

In this sense, the cover provided by the urban tree canopy is very important for the quality of the urban environment and the well-being of the population (NOWAK; CRANE; STEVENS, 2006). Urban trees, with the cover provided by their canopy, have various functions that promote the quality of the environment. The vegetation canopy in urban environments removes atmospheric pollution (NOWAK; CRANE; STEVENS, 2006), reduces air temperature (NOWAK; CRANE; STEVENS, 2006), reduces noise (BOLUND; HUNHAMMAR, 1999), promotes carbon sequestration by reducing carbon dioxide in the atmosphere (NOWAK; CRANE, 2002), reduces energy use in buildings (NOWAK; CRANE, 2002) and also reduces UV radiation on people

(NA et al., 2014). All of these functions can increase their positive impacts the greater the cover provided by urban tree planting in cities.

Urban afforestation offers ecological benefits to its residents, known as ecosystem services (GÓMEZ-BAGGETHUN et al., 2013) and can also be of great importance for the maintenance of urban fauna (TOLEDO, 2006). Ecosystem services can be obtained through ecosystem processes such as climate regulation, air purification, pollination and even erosion control.

The functions of urban afforestation are desirable for the quality of life of city dwellers. This can be extended to university *campuses*, which are large urbanized green areas with a lot of people circulating within them.

A university *campus* can be considered an "urban open space" (GUZZO; CARNEIRO; OLIVEIRA JÙNIOR, 2006). In general, these spaces have large open areas and may have more wooded areas compared to urban centers, as they do not have large densities of buildings and urban infrastructures. Some authors point to the ecological importance and ecosystem services of university *campuses* in peri-urban areas due to their large wooded areas, which can also serve as ecological corridors (COLDING, 2007). Considering this importance, a number of studies have been carried out in universities to analyze and guarantee a quality wooded environment on their *campuses* (BANERJEE et al., 2011; BUENO; XIMENES, 2011; GÜNTHER, 1994; KURIHARA; IMANA-ENCINAS; PAULA, 2005; LEAL; PEDROSA-MACEDO; BIONDI, 2009; MELO; SEVERO, 2007; OLUDUNFE, 2011), but this topic has still not been studied much on university *campuses*.

The Federal University of Sao Carlos (UFSCar) has a non-urbanized area corresponding to 82.32% of the total area, which is mostly used for eucalyptus production and the maintenance of natural vegetation. The predominant types of natural vegetation on the UFSCar Sao Carlos *campus* are wooded savannah, forest savannah and alluvial seasonal forest (riparian vegetation). The importance of the diversity of habitats *on campus* is highlighted, especially in the areas of savannah vegetation and riparian forest, in the surrounding fragments of plateau forests (mesophytic semi-deciduous forests), through the recognition of 46 species of mammals and 212 species of birds (MOTTA- JUNIOR; VASCONCELOS, 1996; MOTTA-JUNIOR et al., 1996).

In relation to its urbanized area, UFSCar's Institutional Development Plan (PDI) provides for the need to provide an urban environment that is in harmony with nature and to encourage the development of research and actions to disseminate knowledge about the environment, the sustainable use of natural resources, environmental preservation and conservation, the minimization of socio-environmental impacts and the adoption of agro-ecological practices. It also foresees the need to provide an urban environment that is in harmony with nature and to draw up a plan for the afforestation of the urbanized areas of the *campus*, prioritizing the planting of native species (UFSCAR, 2013). In this way, formulating and maintaining a tree-planting plan is a very effective way of achieving the desired goal.

The municipality of Sao Carlos also has a Tree Planting Plan (PMSC, 2009) with some technical guidelines to avoid conflicts and make better use of tree planting. This plan contains specifications on distances to be respected from urban infrastructure and even specifications on planting, preventing trees from dying soon after planting or growing with damage and pests. This document also recommends species for planting in urban areas, encouraging the use of species native to the region, but recommending some exotic species due to their adaptation and favorable size for the urban environment.

In this sense, the proper management of urban trees requires management from planting and care throughout their lives until they are removed when they begin to pose a risk of falling.

According to NOWAK et al. (2008), in order to manage urban trees and forests properly, it is essential to have data on this resource, such as the number of trees, the species that make them up, their location, etc. Structural data is fundamental for better planning of urban forestry and for helping to maintain or improve environmental quality, health and well-being in cities. According to this author, the most accurate way of assessing the structure of tree planting is to measure and record each tree individually. A complete census would be ideal, especially for small areas, but meaningful data can be generated with random samples of the environment, mainly with a view to saving resources.

Geographic Information Systems can help with data collection and spatial analysis. This spatialization of the data could improve the reading and visualization of the information, helping with decision-making and management of tree planting. Some studies have already used geoprocessing tools to study urban afforestation (FALCE et al., 2012; LIMA NETO; BIONDI; ARAKI, 2010; OLIVEIRA FILHO; KÜSTER DA SILVA, 2010; SILVA FILHO; COSTA; POLIZEL, 2012; SPADOTTO; DELMANTO JÛNIOR, 2009).

CHAPTER 2

2 OBJECTIVES

2.1 General Objectives

To assist in the planning of urban tree planting on the *campus of* the Federal University of Sao Carlos, providing an informational tool for decision-making in the management of the *campus*.

Carry out an inventory of urban tree planting, create a georeferenced database using the Geographic Information System (GIS), and generate information and guidelines for the management and planning of urban tree planting in the northern area of the *campus of* the Federal University of Sao Carlos (UFSCar), thus providing tools and information for decision-making.

2.2 Specific objectives

- To analyze the richness, diversity and nature of the species (native, exotic and invasive exotic) in the urban forestation of the North Area of UFSCar - Sao Carlos;
- Analyze the relationship between urban forestry and urban infrastructure in the North Area of UFSCar - Sao Carlos, checking for conflicts;
- Analyze the urban vegetation canopy in the North Area of UFSCar - Sao Carlos;

CHAPTER 3

3 MATERIALS AND METHODS

3.1 Study area

The Federal University of Sao Carlos Sao Carlos *campus* (UFSCar - Sao Carlos) is located in Brazil, in the state of Sao Paulo, in the municipality of Sao Carlos, to the north of the city (Figure 1). It is located between the geographic coordinates 21°58 and 22° 00' south latitude and 47° 51' and 47° 52' west longitude, its altitude varies between 815 and 895 m and its climate can be classified according to Koppen as Cwa (tropical with wet summers and dry winters) and Awa (hot, with a well-defined dry period). The average annual temperature is 19.6 °C (MELÂO et al., 2011).

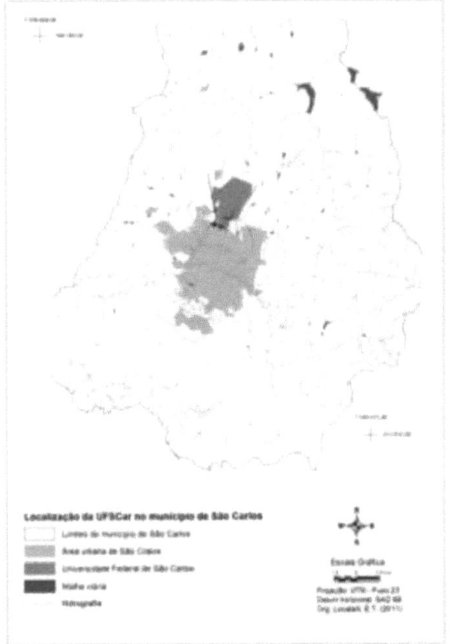

Figure 1: Location of the Federal University of Sao Calos - Sao Carlos. Source: MELÂO, et al. (2011).

The *campus is* divided into two areas, urbanized and non-urbanized. The urbanized area includes all the infrastructure that the university needs, with classrooms, laboratories, administrative buildings, some fragments of natural vegetation and eucalyptus trees, as well as the Monjolinho stream, which divides the university into two areas generically known as the "North Area" and the "South Area". The non-urbanized area also includes some areas of savannah and eucalyptus and some springs (MELÂO et al., 2011). The work was carried out in the urbanized area comprising the North Area of the UFSCar - Sao Carlos *campus*. The study area covers approximately 81.7 hectares (Figure 2).

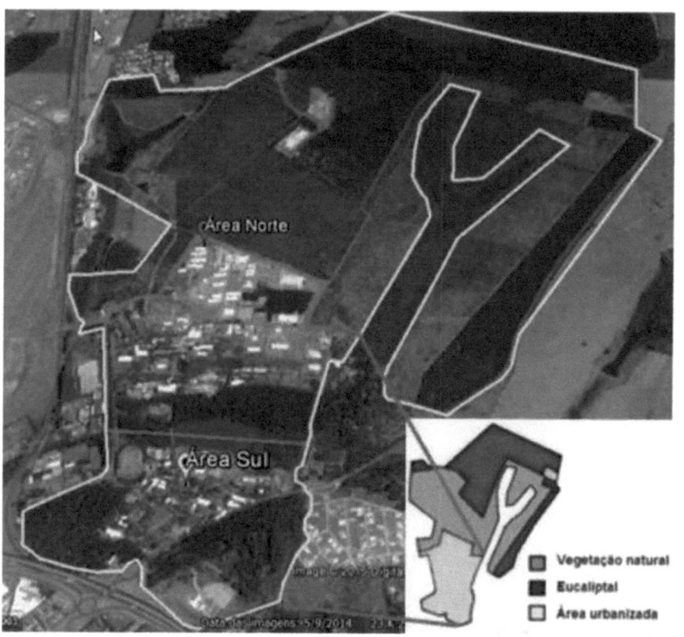

Figure 2: Boundaries, land use, division of the urban area and the study area (North Area) at UFSCar -Sao Carlos. Source: Google Earth and http://www.ufscar.br/~debe/reserva/paginas/ufscar/caracteristicas.htm.

Elaboration: Lessi, B. F. (2014).

The Federal University of Sao Carlos Sao Carlos *campus* (UFSCar-SC) has 26% (162.21 ha) of Legal Reserve Areas (ARL) within its total area (632.42 ha, 22 km perimeter). In addition to the ARL, there are also Permanent Preservation Areas (APP) which occupy 61.77 ha. If added together, the ARL and APP make up a total of 203.54 ha, totaling approximately 32% of areas with native vegetation within the Sao Carlos *campus* (MELÂO et al., 2011). UFSCar also set up a Special Coordination Office for the Environment (CEMA), today transformed into Secretariat for Environmental Management and Sustainability (SGAS), which works on programs to preserve the green areas of the *campuses*, waste management, environmental education and electricity conservation (in conjunction with the University Prefecture). The current SGAS has been dedicated to implementing environmental management and conservation projects on all UFSCar *campuses*.

The urbanized area represents 17.67% of the total *campus* area. This area is used for teaching, research and extension activities. It has a total university population of 13,917 users (MELÂO, 2010).

The total area of the *campus* corresponds to approximately 9.5% of the urban area and 0.5% of the total area of the municipality of Sao Carlos. The *campus* population represents approximately 6% of the population of Sao Carlos, which is 221,950 according to the IBGE (2010 Demographic Census).

3.2 Information gathering

The inventory and preparation of the georeferenced database was carried out in stages, divided as follows:

- Stage 1: Characterization of the study area, the North Area of the urbanized area of UFSCar - Sao Carlos, and bibliographic survey of research related to the topic of urban afforestation;
- Stage 2: Beginning of the inventory with the identification of urban tree species;
- Stage 3: Detailed technical analysis of urban afforestation;
- Stage 4: Setting up the database in the GIS.
- Stage 5: Digitization of the urban vegetation canopy.

To carry out the activities, SGAS and the Physical Development Office (EDF) collaborated to provide guidance and information and equipment. The EDF provided the General Map of UFSCar - Sao Carlos, and SGAS provided equipment and infrastructure for field and laboratory work. This formed a partnership between SGAS,

EDF and the Postgraduate Program in Ecology and Natural Resources (PPG-ERN).

In order to make the fieldwork even more complete, undergraduate students from the Environmental Analysis and Management course at the *campus* itself were selected. These students received supervised training to standardize the methodology so that they could then help collect the data.

3.2.1 Stage 1 - GIS digitization of the study area

The General Map of the UFSCar Sao Carlos Campus (2013) (Figure 3) was digitized in the Geographic Information System (GIS) (Quantum GIS - QGIS software) using a planialtimetric survey provided by the Physical Development Office (EDF). The General Map of the UFSCar Sao Carlos Campus (2013) contains all the university's infrastructure, such as streets, squares, blocks, buildings, urban vegetation fragments, urban trees *on campus*, among others, which were exported and digitized in the GIS. During the field activities, some other individuals from the urban forest were included that were not on the General Map provided by the EDF.

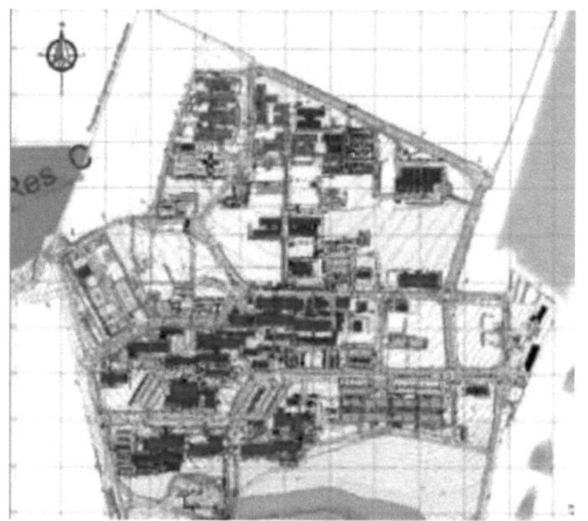

Figura 3: Cutout of the General Map of UFSCar - Sao Carlos (2013) in the North Area of urbanized area of UFSCar - Sao Carlos. Source: General Map of UFSCar - Sao Carlos (2013). Elaboration: Lessi, B. F. (2014).

After exporting the data from the General Map (UFSCAR, 2013) to the GIS, the information was separated into layers: one layer with the digitalization of the buildings, one for the asphalt sidewalk, one for the sidewalks and tiled sidewalks, one for the grassy areas, one for the patches of dense vegetation found on *campus* and one for the urban tree cover. The layer for urban tree information contained all the mapped individuals, represented by a point with their geographical coordinates.

3.2.2 Stage 2 - Field: Area reconnaissance and species identification

3.2.2.1 Sampling methodology

With the urban forest *already* mapped in the previous stage, a fixed identification number was generated for each tree in the GIS. After this procedure, a GRID was created in the GIS, dividing the study area into 100 m x 100 m plots (10000 m^2) to organize the field collections (Figure 4). Small maps of the areas to be analyzed were then created. These small maps contained the layers of infrastructure information digitized in the previous stage, such as streets, buildings, etc. in order to facilitate the exact location of the area and the individual trees. The maps also contained the numbering of the tree points, which served as a guide for the field analysis of each individual (Figure 5).

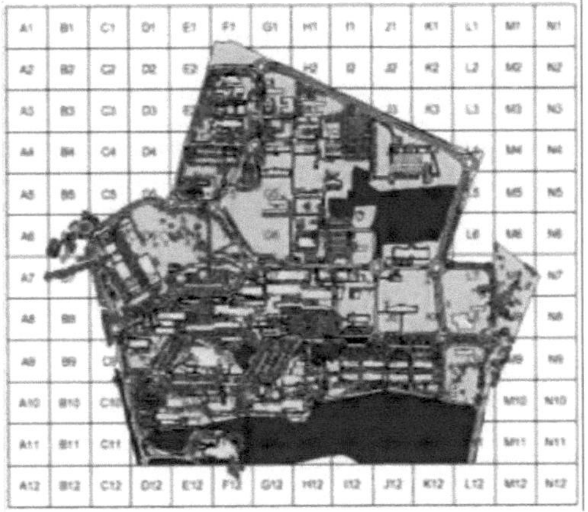

Figura 4: Representation of the GRID of 100m x 100m plots dividing the study area in the GIS (North Area of UFSCar - São Carlos). Source: EDF UFSCar (2013) and Lessi B. F. Elaboration: Lessi, B. F. (2014).

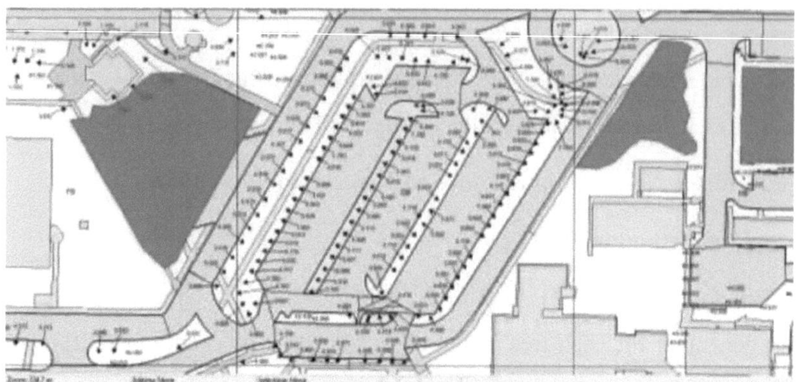

Figure 5: Small map taken to the field, printed, with the numbering of the individual trees and the urban infrastructures mapped for easy location of the area and the individual. In the center of the plot is its reference, in this case G9, in the North Area of UFSCar - Sao Carlos. Source: EDF UFSCar (2013). Elaboration: Lessi, B. F. (2014).

The data collection began in the field, based on the maps formed with the trees to be identified and analyzed. To do this, a printed spreadsheet was taken to the field with all the information to be collected that would make up the GIS database.

The inventory was carried out on as many individuals as possible over a period of one year. The activities began with an analysis of the individuals around Rua da Biblioteca, which is the main road in UFSCar's North Area, and then proceeded randomly, drawing plots from the GRID generated using the GIS.

3.2.2.2 Species identification

At this stage, species were identified from the components of urban tree planting and fruit species were selected:

Species identification: Individuals were identified to species level. Individuals were identified in the field using identification catalogs (Lorenzi H, 2003; 2008; 2009a; 2009b) and when identification in the field was not possible, samples of leaves, fruits and flowers were collected and taken to the laboratory for drying and identification. The natural occurrence of each species was verified according to the literature (Lorenzi H, 2003; 2008; 2009a; 2009b) (native: occurs naturally in phytophysiognomies found within Brazilian territory; local native: occurs naturally in phytophysiognomies in the state of Sao Paulo and exotic: occurs naturally in regions outside Brazilian territory).

Frutiferous species: The classification of species for this category was based on the diet of human beings, the type of dispersal (zoochoric) and their attractiveness to local fauna according to Frisch and Frisch (2005) and Kuhlmann (2012); Two codes were considered for this identification: code 0 (Zero): non-frutiferous; code 1: frutiferous.

3.2.3Stage 3 - Field: Technical analysis

The methodology for data collection was based on that described by ALBRECHT (1998), with some modifications to better fit the environment studied. For the analysis in the field, a spreadsheet was used which contained all the information to be collected from the observation of each individual (Figure 6).

Arborização UFSCar São Carlos

Nº	ESPÉCIE	H	HB	CAP	DC	C	R	CE	CA	CT	CF	CS	CP	CCT	CI	CO	MF	CT	AL	PL	CFITO	DFV	DFP	DFT	DFN	DFO	FRUT	NE	OC	IA	AF	AR

Figure 6: Field worksheet for data collection. Source: Based on and modified from ALBRECHT (1998). Prepared by: Lessi, B. F. (2014).

3.2.3.1 Size

H - Height: The height of the trees (from the ground to the highest part of the crown) was checked using a hypsometer, an instrument capable of measuring by triangulation.

HB - Height of the 1st bifurcation: The height of the stem was measured, i.e. from the plant's insertion into the ground to the first insertion of branches from its trunk (bifurcation);

CAP - Circumference at Breast Height: The diameter of the trunk at approximately one and a half meters from the ground was measured. When there was a bifurcation right at the beginning of the trunk, the basal diameter of the plant was considered, and if the plant had several bifurcations from the ground, all the bifurcations were measured approximately one and a half meters from the ground and then all the measurements were added together, calculating an average;

3.2.3.2 Presence of conflict

Possible conflicts between trees and urban structures or functions were assessed. Conflicts were considered to be any situation in which an individual tree was disturbing, inconveniencing or damaging urban infrastructure (Code 0 (Zero) - No conflict; Code 1 - Conflict present);

CE - Buildings: every time a tree came into contact with a building or had its branches above it, it was considered a conflict with buildings. For example, a tree with a very large crown with branches touching the windows of buildings, causing a risk of glass breakage or damage to the structure of these windows. Conflict with buildings was also considered when an individual tree was in danger of falling on nearby buildings because it had phytosanitary problems or if it had bent due to heavy rain or wind;

CF - Power grid: Whenever the tree was leaning against the power line, for example, wide and tall canopies that grew in the middle of the power line, tall trees that could fall on the power line causing damage to the power line, such as short circuits, falling poles;

CP - Sidewalk: When the tree is in the way of pedestrians crossing sidewalks, either because of its large crown or because it is leaning towards the sidewalk;

CI - Lighting: When the tree is obstructing the lighting of a street lamp;

CR - Roots: The situations found between the roots and the urban environment were characterized: Code 1 - Roots evident and damaging the sidewalk; Code 2 - Roots underground, but already damaging the sidewalk; Code 3 - Roots underground and not damaging the sidewalk; Code 4 - Roots evident but not damaging the sidewalk;

3.2.3.3 PFD - Physical damage caused by pruning

Whether or not there has been physical damage after some type of pruning, such as trunks with damaged development, crooked or malformed trunks, poor healing or the presence of diseases caused by delayed healing. It was assumed that any pruning causes injury and physical damage to the individual, since pruning is not a natural process for the plant. Code 0 (Zero) No damage; Code 1: Presence of apparent damage;

3.2.3.4 Need for pruning

The need for management was assessed by analyzing the information on conflicts and the individual's physical condition, as described above. The following criteria were used: corrective pruning (MASCARÓ; MASCARÓ, 2010) to eliminate and/or mitigate conflicts; conductive pruning (MASCARÓ; MASCARÓ, 2010) to prevent conflicts and cleaning or maintenance pruning (MASCARÓ; MASCARÓ, 2010) to eliminate dry or diseased branches. All of this was considered to be in need of pruning intervention (Code 0 (Zero): No pruning required; Code 1: Pruning required);

3.2.3.5 Positioning and permeable planting area

The positioning in relation to the street curb and the quality in relation to the free permeable area at the base of the stem of each individual were evaluated.

MF - distance from the curb: The distance the individual is planted from the curb, i.e. from the center of the base of the trunk close to the ground to the center of the curb, was taken into account;

CT - Building: The distance from the building closest to the individual was taken into account. Distances of up to 20m were measured; distances greater than 20m were considered to be greater than 20 meters;

AL - Permeable free area: The permeable area of each individual, as well as the size of the hole opened for planting, is very important for the penetration of water, oxygen, salts and organic matter necessary for the survival of the plant. were evaluated with codes from 1 to 4, with 1 for free areas considered good, with 3 or more square meters, code 2 for free areas between 1 and 3 m^2, code 3 for areas below 1 m2 and 4 for individuals without any free area, with paving or cobbles up to the neck of the plant. These standards were based on those recommended in the literature (MATOS; QUEIROZ, 2009; PSP, 2005) for healthy urban tree planting.

3.2.3.6 General state of afforestation

The general condition of each individual was determined according to the following factors:

CFITO - Phytosanitary control: The need for phytosanitary control was assessed when the presence of visible pests and diseases was detected, causing damage to the development of the individual tree. Phytosanitary control must be analyzed on a case-by-case basis in the presence of a qualified professional, and it is extremely hasty and erroneous to apply treatments without analysis by the correct professional. Code 0 (Zero): No need for phytosanitary control, apparently the individual is in good condition and there is no appearance of pests or diseases; Code 1: Apparent need for phytosanitary control, the assessment of a professional is necessary to verify the real need and the correct treatment for a given pest or disease;

IA - Integrity: The general state of the individual, its integrity, was assessed with codes from 1 to 3, where code 1 for good integrity, code 2 for average integrity and code 3 for poor integrity. This took into account the development of the trunk and canopy, the need for phytosanitary control, apparent physical damage and injuries;

GF - Degree of functionality: The functionality of each individual tree was defined according to its location and the use of its services by the community. Code 1 was considered to be the minimum for all trees, since all trees have a minimum function in an environment just by existing. Code 2 was given to trees that are visited a lot, such as near sidewalks, and parking lots where the flow of people is greater. Code 3 was given to individuals that were visited a lot by people for leisure, or near canteens where people could sit in their shade, as well as being near sidewalks or places where people flow;

AR - Risk analysis: The **risk** of the tree falling was assessed, as well as the risk of it causing material damage or posing a risk to people. Risks were considered in codes from 0 to 4, taking into account the conflicts that the tree presents (the more conflicts, the greater the risk of damage), the integrity of the individual (general condition, so the worse its condition, the greater the risk), the need for phytosanitary controls (the risk increases with the need for phytosanitary control), the presence of physical damage (the presence of serious physical damage increases the risk), proximity to buildings (the closer to buildings the greater the risk of material damage), the degree of functionality, which considers how important the tree is to the environment (the greater the functionality, the greater the risk of causing harm to people in a possible fall). Risk codes between 2 and 3 will require an assessment by a qualified professional. Code 3 is used for delicate cases that require urgent intervention; Code 2 for cases that require special attention. When the individual is dry or apparently dead, code 4 is used, indicating the need for more urgent intervention. Taking into account all the stress that the urban environment brings to its vegetation (pollution, soil compaction, lack of nutrients, etc.), risk 1 was considered to be the minimum risk that a tree can present, i.e. it will always present a risk being in an urban environment.

3.2.4Stage 4 - GIS database and data analysis

The urban trees, which were represented by dots, were given an identifying number, the reference for locating and digitizing the field data for each individual. Then, in addition to the numerical identifier, a database was set up with the information defined to be collected during the field analysis (Figure 7). This database in the GIS was filled in after the field stage, where the information was collected. Once the information had been collected in the field, it was entered into a spreadsheet and then exported to the GIS to make up the georeferenced database. We calculated species dominance indices (D), equability indices (J) and the Shannon-Weaver diversity index, which is based on the proportional abundance of species, assuming that all the species present have been identified. The indices were calculated using the statistical program Past V. 2.17b (HAMMER; HARPER; RYAN, 2001).

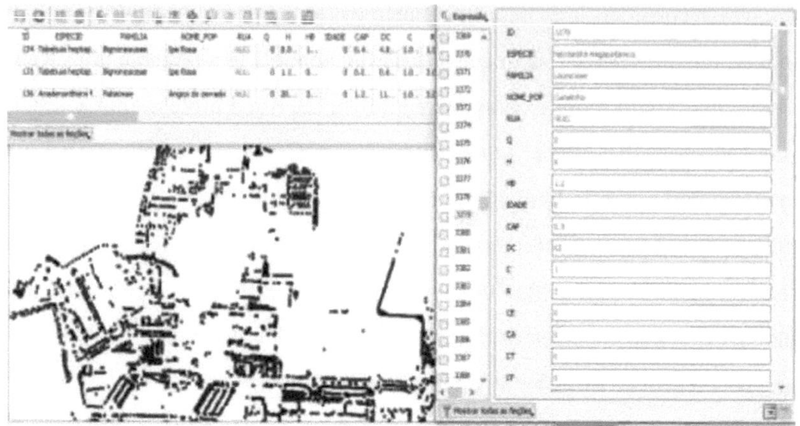

Figure 7: Database for urban tree inventory in GIS. In the center of the figure are the mapped points of the individuals that make up the urban afforestation of the North Area of the UFSCar - Sao Carlos *campus*. To the right of the figure is a demonstration of the database accessed by individual (in this case individual code number 3379), and at the top is the beginning of the complete database, with all the individuals and their information. Source: EDF UFSCar (2013). Elaboration: Lessi, B. F. (2014).

3.2.4.1 Stage 5 - Land use and occupation and the urban vegetation canopy

The use and occupation analysis was carried out by digitizing the information from the General Map of UFSCar - São Carlos (2013). For the analysis of the urban vegetation canopy, the canopy of all the urban trees and the vegetation fragments (forestry and native vegetation) present in the study area were considered. Polygons were digitized over the tops of trees (alone or in groups), according to Nowak (1996), which is a good methodology for estimating tree cover. Digitization was carried out using Google Earth software (2013) and the polygons were exported to the QGIS program for editing, adjustments and data analysis. Then, by overlaying this layer of information on tree cover with the other layers of land use and occupation, it was possible to make a brief analysis of the cover of this vegetation on the streets and sidewalks.

CHAPTER 4

4 RESULTS AND DISCUSSION

4.1 Field: Area reconnaissance and species identification

Initially, after digitizing the General Map of UFSCar - Sao Carlos (2013), 5871 points were counted, which represented the individuals of the urban arborization of the entire *campus,* with 3869 individuals present in the North Area of the *campus*, the study area.

During the field activities in the North Area of the *campus*, 324 individuals were found that were not initially recorded. These individuals were mapped and included in the database, which gave a total of 4,193 individuals in the study area. As a result, the urban tree cover of the study area represents 67.6% of the tree cover of the entire UFSCar - Sao Carlos *campus*.

During the study, 3020 individuals were inventoried (Figure 8), which corresponds to 72% of the urban afforestation in the study area and 51.4% of the total urban afforestation at UFSCar.

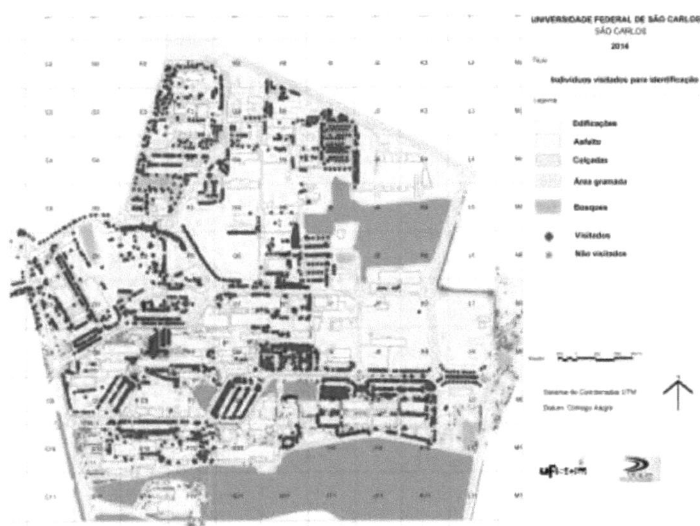

Figure 8: Individuals visited during the inventory of urban trees in the North Area of UFSCar - Sao Carlos, SP, totaling 3020 individuals. Source: EDF UFSCar (2013) . Elaboration: Lessi, B. F. (2014).

Thus, of the 3020 individuals inventoried, 2548 individuals were identified as species, 302 individuals did not have their species identified and were recorded as "ID" (Identify). There were 148 points recorded on the General Map (UFSCar 2013) where the individuals were no longer to be found, so they were considered to be suppressed. Finally, 22 dry or dead individuals were found (Table 1).

The urban afforestation in the northern area of UFSCar - Sao Carlos is made up of 130 species, 107 genera and 46 families. Among these species, there are 51 local native species, 13 species native to other regions of Brazil, 66 exotic species and 6 exotic species with invasive potential. This community has different morphological characteristics such as growth and size (MASCARÓ; MASCARÓ, 2010), and is therefore composed of 102 tree species with 2160 individuals, 21 shrub species with 166 individuals and 7 palm species with 222 individuals, showing a predominance of tree species.

According to Aguirre Junior (2006), medium and large tree species bring more environmental benefits to the urban environment than shrub species. Planting shrubs should therefore not be avoided, but trees should not be replaced by shrubs simply because they are easy to manage and adapt to the urban environment due to their small size.

Table 1: Species recorded in the survey of urban trees in the northern area of UFSCar Family; Species (*-Native to the region; §-Native to Brazil; †-Exotic; ∞-Exotic invasive. □-Arboreal; ◊-Shrub; --Palm); Common name; NI- Number of individuals; Freq.- Frequency of recording).

Family Species	Common name	NI	Frequency
ACANTHACEAE			
Sanchezia oblanga Ruiz & Pav † ◊	Sanquesia	1	0,03
AGAVAcEAE			
Furcraea foetida (L.) Haw. * ◊	Piteira	48	1,59
ANAcARDIAcEAE			
Anacardium occidentale L. * □	cashew tree	4	0,13
Mangifera indica L. † □	Hose	42	1,39
Schinus molle L. § □	Aroeira Salsa	194	6,42
Schinus terebinthifolius Raddi. * □	Red mastic	7	0,23
Spondiaspurpurea L. † □	Seriguela	1	0,03
ANNONAcEAE			
Annona squamosa L. * □	Anona	4	0,13
APOcYNAcEAE			
Piumeria rubra L. † □	Mango jasmine	17	0,56
Thevetia thevetionoides (Kunth) K. Schum † ◊	Napoleon hat	16	0,53
ARAUcARIAcEAE			
Araucaria angustifolia Raddi § □	Araucaria	13	0,43
Araucaria calimnaris (J. R. Forst.) Hook † □	Christmas tree	1	0,03
AREcAcEAE			
Archontophoenix cunninghamiana H. Wendl. &. Drude † -	Seafortia	17	0,56

Keep going...

Table 1: Species recorded in the survey of urban trees in the northern area of UFSCar Family; Species (*-Native to the Region; §-Native to Brazil; †-Exotic; ∞-Invasive exotic. □- Arboreal; ◊-Shrub; --Palm); Common name; NI- Number of Individuals; Freq.- Record frequency)

Family Species	Common name	NI	Frequency
Caryota urens L. †-	Fishtail palm	2	0,07
Chamaerops humilis L. †-	Palm do Mediterranean	2	0,07
Dypsis lutescens (H. Wendl.) Beentje & J. Dransf. †-	Areca bamboo	28	0,93
Phoenix roebelenii O'Brien †-	Phoenix	27	0,89
Roystonea oleracea L. H. Bailey	Imperial Palm	14	0,46
Syagrus romanzoffiana (Cham.) Glassman * -	Jerivà	132	4,37
ASPARAGACEAE			
Agave americana L.◊	Blue firecracker	1	0,03
Dracaenafragrans (L.) Ker Gawl. † ◊	Waterwood	4	0,13
Dracaena marginata Lam. † ◊	Dracena rainbow	6	0,20
Yucca elephantipes Regel. † ◊	Elephant ling	6	0,20
ASTERACEAE			
Baccharis dracunculifolia DC. * ◊	Broom	3	0,10
BIGNONIACEAE			
Espathodea nilotica Seem. ∞ □	Swordfish	3	0,10
Handroanthus ochraceus (Cham.) Mattos * □	Yellow Ipe	52	1,72
Jacarandà brasiliana (Lam.) Pers. * □	Jacaranda mouth frog	31	1,03
Jacaranda macrantha Cham. * □	Caroba	2	0,07
Jacaranda micrantha Cham. § □	Caroba	1	0,03
Jacaranda mimosifolia D. Don□	Jacarandà mimoso	1	0,03
Tabebuia heptaphylla (Vell.) Toledo * □	Pink Ipê	175	5,79
Tabebuiapentaphylla Hemsl. □	Ipê de El Salvador	11	0,36
Tabebuia roseoalba (Ridl.) Sandwith * □	White Ipe	134	4,44
Tecoma stans (L.) Juss. Ex Kunth□	Garden Ipe	10	0,33
BOMBACACEAE			
Bombax malabaricum DC. □	Paineira da india	2	0,07
CARICACEAE			
Carica papaya L.□	Papaya	1	0,03
CLUSIACEAE			
Garcinia brasiliensis Mart. § □	Mangostao	1	0,03
COMBRETACEAE			
Terminalia catappa L. ∞ □	Seven hearts	1	0,03

Keep going...

Table 1: Species recorded in the survey of urban trees in the northern area of UFSCar Family; Species (*-Native to the Region; §-Native to Brazil; †-Exotic; ∞-Invasive exotic. □- Arboreal; ◊-Shrub; --Palm); Common name; NI- Number of Individuals; Freq.- Record frequency)

Family Species	Common name	NI	Frequency
CRYSOBALANACEAE			
Licania tomentosa (Benth) § □	Oiti	46	1,52
CUPRESSACEAE			
Callitropsis macrocarpa (Hartw. Ex Gordon) D.P. Little◊	Monterrey cypress	14	0,46
Cupressus Iisitanica Mill□	Cedar	28	0,93
CYCADACEAE			
Cycas revoluta Thunb. ◊	Cica	9	0,30
DILLENIACEAE			
Dillenia indica Blanco□	Money tree	2	0,07
ERICACEAE			
Rhododendron sımsıı Planch◊	Azalea	14	0,46
EUPHORBIACEAE			
Codiaeum variegatum (L.) Rumph. Ex A. Juss◊	Croton	2	0,07
Croton floribundus Spreng. * □	Capixingui	9	0,30
Euphorbia Ieucocephala Lotsy◊	Mountain Snow	2	0,07
FABACEAE			
Acacia podalyraefolia A. Cunn. Ex G. Don † □	Acacia mimosa	3	0,10
Albizia polycephala (Benth.) Killip * □	White Angus	14	0,46
Albizia niopoides (Spruce ex Benth.) Burkart * □	Dry flour	7	0,23
Anadenanthera falcata (Benth.) Speg * □	Angico do cerrado	16	0,53
Anadenanthera macrocarpa (Benth.) Brenan * □	Red Angle	11	0,36
Anadenantherapavonina L.□	Pavao Eye	22	0,73
Bauhinia forficata Link * □	Cow's foot	2	0,07
Bauhinia variegata L.□	Cow's foot	15	0,50
Caesalpinia echinata Lam. § □	Pau Brasil	3	0,10
Caesalpinia férrea Mart.* □	Ironwood	98	3,25
Caesalpinia pluviosa DC. * □	Sibipiruna	221	7,32
Caesalpiniapulcherrima (L.) Sw◊	Mini Flamboyant	12	0,40
Calliandra brevipes Benth. § ◊	Caliandra	1	0,03
Calliandra tweedii Benth. § ◊	Red Caliandra	1	0,03
Cassia fistula L.□	Cassia imperial	1	0,03
Cassia grandis L. § □	Cassia rosa	5	0,17

Keep going...

Table 1: Species recorded in the survey of urban trees in the northern area of UFSCar Family; Species (*-Native to the Region; §-Native to Brazil; †-Exotic; ∞-Invasive exotic. □- Arboreal; ◊-Shrub; --Palm); Common name; NI- Number of Individuals; Freq.- Record frequency)

Family Species	Common name	NI	Frequency
Cojoba sophorocarpa (Benth.) Britton & Rose § □	Siraricito	15	0,50
Dalbergia miscolobium Benth. * □	Jacarandà do cerrado	14	0,46
Delonix regia (Bojer ex Hook.) Ralf□	Flamboyant	8	0,26
Enterolobium confortisiliquum (Vell.) Morong * □	Timburi	28	0,93
Erythrina verna Vell. * □	Mulungu	2	0,07
Hymenaea courbaril L. * □	Jatobà	3	0,10
Hymenaea stigonocarpa Mart. Ex Hayne * □	J atobà-do-cerrado	1	0,03
Inga vera subsp. Affinis (DC.) nT.D. Penn. * □	Ingà-do-brejo	10	0,33
Machaerium Acutifolium Vogel * □	Field jacaranda	12	0,40
Peltophorum dubium (Spreng.) Taub. * □	Canafistula	32	1,06
Pterocarpus violaceus Vogel * □	Aldrago	3	0,10
Pterogyne nitens Tul. * □	Wild peanut	1	0,03
Senna pendula (Humb.& Bonpl. Ex Willd.) * ◊	Sticky straw	5	0,17
Senna Siamea (Lam.) H.S. Irwin & R. C. Barneby□	Siamese Cassia	43	1,42
Stryphnodendron adstringens (Mart.) Coville * □	Real Barbatimao	12	0,40
Tamarindus indica L.□	Tamarind	1	0,03
LAURACEAE			
Nectandra megapotamica (Spreng.) Mez * □	Canelinha	55	1,82
Persea americana Mill□	Avocado tree	13	0,43
LYTHRACEAE			
Lafoensia glyptocarpa Koehne § □	Mirindiba-rosa	12	0,40
lagerstroemia indica L.□	Reseda	35	1,16
Punica granatum L.□	Rome	2	0,07
MAGNOLIACEAE			
Michelia champaca L.□	Magnolia	32	1,06
MALPIGHIACEAE			
Malpighia glabra L.□	Acerola	8	0,26
MALVACEAE			
Ceiba speciosa (A. St.-Hill.) Ravenna * □	Paineira Rosa	3	0,10

Keep going...

Table 1: Species recorded in the survey of urban trees in the northern area of UFSCar Family; Species (*-Native to the Region; §-Native to Brazil; f-Exotic; ∞- Invasive exotic. □- Arboreal; ◊- Shrub; --Palm); Common name; NI- Number of Individuals; Freq.- Record frequency)

Family Species	Common name	NI	Frequency
Eriotheca gracilipes (K. Schum.) A. Robyns * □	Paineira-do-cerrado	8	0,26
Guazuma crinita Mart.* □	Mutamba	13	0,43
Guazuma ulmifolia Lam. * □	Mutambo	11	0,36
Sterculia chicha A. St.-Hill. Ex Turpin □	Meat	4	0,13
MELASTOMATACEAE			
Tibouchina granulosa (Desr.) Cogn * □	Lenten tree	84	2,78
Tibouchina mutabilis (Vell.) Cogn § □	Manacà da Serra	1	0,03
MELIACEAE			
Cedrela fissilis Vell * □	Pink Cedar	4	0,13
Guarea guidonia (L.) Sleumer * □	Sailor	2	0,07
Melia uzedarach L.□	Sinamomo	1	0,03
Swietenia macrophylla king□	Mahogany	1	0,03
MORACEAE			
Artocarpus heterophyllus Lam. □	Jaqueira	3	0,10
Ficus benjamina L.□	Ficus	11	0,36
Morus nigra L.□	Amoreira	23	0,76
MUNTINGIACEAE			
Muntingia calabura L.□	Calabura	2	0,07
MYRTACEAE			
Calistemon citrinus (Curtis) Skeels□	Imperial Calistemon	1	0,03
Calistemon viminalis (Sol. Ex Gaertn.) G. Don□	Bottle brush	3	0,10
Eucaliptuspilularis Sm.□	Eucalyptus	57	1,89
Eugenia pyriformis Camb. * □	Grapefruit	5	0,17
Eugenia uniflora L. * □	Pitangueira	36	1,19
Myrcia bella Cambess * □	Mircia	1	0,03
Myrciaria jaboticaba (Vell.) O. Berg * □	Jaboticaba tree	3	0,10
Bougainvillea glabra Choisy * ◊	Spring	3	0,10
Psidium guajava L. * □	Guava tree	45	1,49
Syzygium cumini (L.) Skeels ∞ □	Jambolao	26	0,86
NYCTAGINACEAE NYSSACEAE			
Camptotheca acuminata Decne. □	Happy tree	4	0,13
OLEACEAE			
Ligustrum lucidum W. T. Aiton ∞ □	Tailor	156	5,17

Keep going...

Table 1: Species recorded in the survey of urban trees in the northern area of UFSCar Family; Species (*-Native to the Region; §-Native to Brazil; f-Exotic; ∞- Invasive exotic. □- Arboreal; ◊-Shrub; --Palm); Common name; NI- Number of Individuals; Freq.- Record frequency)

Family Species	Common name	NI	Frequency
PINACEAE			
Pinus elliotti Engelm. ∞ □	Pins	5	0,17
PLATANACEAE			
Platanus orientalis L.□	Plàtano	1	0,03
POLYGONACEAE			
Triplaris caracasana Cham. □	stick ant of caracas	7	0,23
PROTEACEAE			
Grevillea banksii R. Br. ◊	Garden sprout	5	0,17
RHAMNACEAE			
Colubrina glandulosa Perkins * □	Sobrasil	1	0,03
ROSACEAE			
Eriobotryajapônica (Thunb.) Lindi□ ◊	Loquat	38	1,26
Prunus persica (L.) Batsch□	Peach tree	2	0,07
RUBIACEAE			
Calycophyllum spruceanum (Benth.) Hook. f. ex K. Schum. § □	Pau Mulato	1	0,03
Coffea arabica L.◊	coffee	3	0,10
Genipa americana L. § □	Jenipapo	4	0,13
RUTACEAE			
Citrus limon L.□	Limoeiro	13	0,43
Murrayapaniculata (L.) Jack□	False myrtle	12	0,40
Zanthoxylum roipholium Lam. * □	Clog	7	0,23
STYRACACEAE			
Styrax ferrugineus L. * □	Cerrado orange	1	0,03
URTICACEAE			
Cecropia pachystachya Trecul * □	Embaùba	6	0,20
VERBENACEAE			
Duranta repens L. § ◊	Drop of gold	10	0,33
VOCHYSIACEAE			
Qualea grandiflora Mart. * □	Earthwood	1	0,03
UNIDENTIFIED			
Unidentified (ID)	-	302	10,00
Dry/Dead	-	22	0,73
Suppressed	-	148	4,90

A total of 207 seedlings (newly planted individuals) were found among the trees identified, of which 25 are unidentified species, 121 are local native species, 5 are native to other regions of Brazil and 56 are exotic species. This survey reveals a trend towards planting native species, thus complying with the recommendations of the 2013 IDP. The Sao Carlos Urban Tree Planting Plan also encourages the planting of native species in its guidelines, as well as refuting the use of invasive exotic species in urban tree planting, which also justifies the removal of existing specimens (PMSC, 2009).

A greenhouse for growing seedlings was recently built on *campus*, which is very important for the future planning of the site's afforestation. This initiative makes it possible to produce seedlings that would be used in new plantings. In addition, with the effective presence of cerrado areas within the *campus*, seed collection can be carried out without difficulty. Cultivation can reduce planting costs and the plants can be cultivated for longer in order to grow them to the ideal size for better survival and adaptation to the urban environment.

Considering that most of the urban forest is man-made vegetation, with only a few trees left over from the urbanization process or from regrowth and natural seed dispersal, one might think that every urban environment could have roughly the same pattern of species richness, even if these urban environments have different cultural and environmental influences. Many studies reveal a predominance of a few species in these environments (SILVA FILHO; BORTOLETO, 2005) and also the presence of the same species in several cities (SILVA FILHO; BORTOLETO, 2005). As it was an old farm with agriculture (MELÂO et al., 2011), its urbanization process developed in an area with few native trees, so the vast majority of its trees were planted along with the urbanization process.

By way of comparison, we selected some studies with similar numbers of individuals sampled in urban environments, in order to get an idea of the richness found in these places and compare it with that found in this study. Thus, the species richness found in this survey was quite high, with 130 species identified for 2548 individuals inventoried. For example, Silva (2005) in an inventory of 2,551 trees in a neighborhood in Americana (SP) found a richness of 76 species, (Couto 2006) working in a neighborhood in Rio de Janeiro (RJ) found a richness of 80 species from a sample of 1,745 tree individuals. Cadorin (2008) found 55 species in a sample of 2036 trees in a neighborhood of Pato Branco (PR). Stranghetti (2010) identified 67 species in 2640 individuals in Uchôa (SP). Sucomine (2010) found 103 species in 2626 individuals in the center of Sao Carlos (SP). Andreatta (2011) found 95 species in 2465 individuals sampled in Santa Maria (RS). Rossatto (2008) found 54 species with 1915 individuals sampled in the city of Assis (SP).

Analyzing surveys carried out at other universities, Gracioli (2011), evaluating the arborization of the University *Campus of* the Federal University of Santa Maria (RS), found a richness of 75 species in a sample of 1,270 tree individuals. Leal (2009) at *Campus* III of the Federal University of Paraná identified 178 species in a survey of 5034 individuals, including trees, shrubs and herbaceous plants. Faleiro (2007) surveyed 965 trees and found 56 species on the Umuarama *Campus of* the Federal University of Uberlândia, which was considered to be a high level of diversity. At the University of Brasilia, Kurihara and Encinas (2003) found 156 species in a survey of 5,011 trees, which was also considered to be high diversity by the author himself. In Passo Fundo (RS), on the *campus of* the University of Passo Fundo (MELO; SEVERO, 2007), they found 108 species in a sample of more than 3000 trees. UFSCar shows a richness close to that found on other *campuses*, regardless of the number of individuals sampled.

Among the species identified, the most abundant was *Caesalpinia pluviosa* (Sibipiruna) with 221 individuals, which represents 8.6% of the total. Other species had significant numbers of individuals, as can be seen in Table 1. With regard to Genera and Families, the most abundant were the *Caesalpinia* genus with 334 representatives and 13.8% and the Fabaceae Family with 32 species and 622 individuals, representing 24.4% of the total.

The diversity of species in urban tree planting is very important for reducing the incidence of pests and diseases. A high density of a single species is inadvisable, as the susceptibility of large homogeneous populations to pests and diseases can increase (SANTAMOUR, 1990). In this sense, Santamour (1990) describes an urban model suggesting a safe diversity against attacks by insects and diseases, which consists of no more than 10% of the same species, a maximum of 20% of the same genus and 30% of the same family. In the northern area of the UFSCar *campus*, the highest species frequencies are no more than 8.6%. With regard to the frequency of genera and families, the highest frequencies are 13.8% in the genus *Tabebuia* and 24.4% in the family Bignoniaceae, which is lower than that proposed by Santamour (1990). Figure 9 shows the predominance of species with smaller numbers of individuals, up to 50, and only 11 species with more than 50 individuals, which represent 53.2% of the trees.

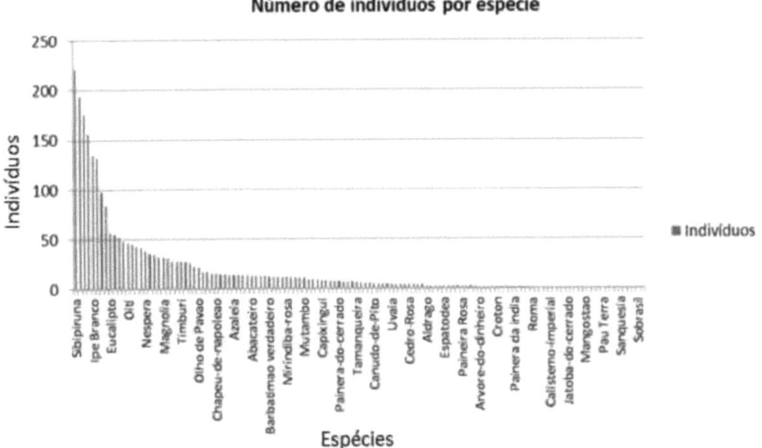

Figure 9: Number of individuals per species in descending order found in the urban tree inventory of the North Area of UFSCar - Sao Carlos, SP. Source: Lessi, B. F. (2014). Elaboration: Lessi, B. F. (2014).

The analysis using the Georeferenced Database shows that the species with the highest frequencies have individuals grouped together (Figure 10), and therefore, although their frequency in the study area as a whole is below 10%, their susceptibility to the occurrence of diseases can still be high, as the proximity between individuals of the same species can facilitate the transmission of pests and diseases.

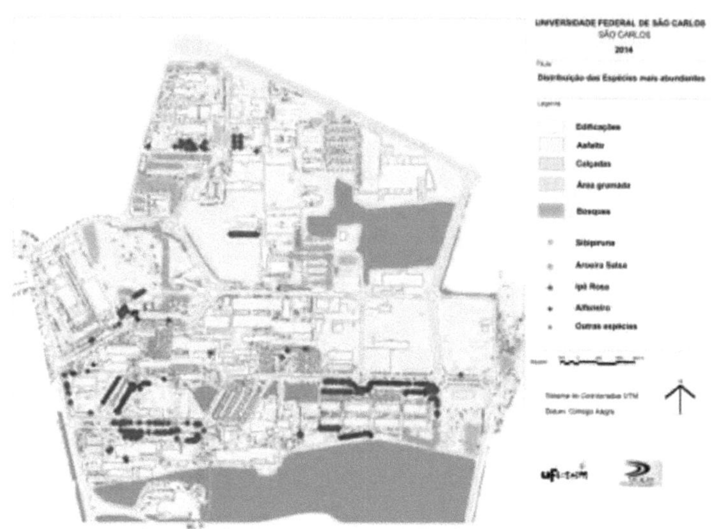

Figure 10: Distribution of the most abundant species found in the inventory of urban trees in the North Area of UFSCar - Sao Carlos, SP Source: EDF UFSCar (2013) . Elaboration: Lessi, B. F. (2014). .

The Shannon-Weiner diversity index (H'), which takes into account species richness and abundance, was 3.9 in the study area. Other surveys carried out in urban areas have calculated diversity. Meneghetti, (2003) in Santos, found a diversity of 2.63; Sucomine (2009) found a diversity of 3.18 in the center of Sao Carlos; Bortoleto (2004) found 3.9 in Aguas de Sao Pedro and Maza (2002), in Santiago, Chile, found a diversity of 3.24 in areas of public afforestation and 4.01 in private areas.

The low dominance of species found in this community surveyed *on campus*, which was 0.03, contributes to a more diverse community with high equability (J = 0.8, Pilou index). Low dominance means that none, or a group of species, has a much higher abundance than the other species, which makes the community more balanced and more equable. In an ideal situation, with maximum equability (J = 1), a community would have all species with the same number of individuals, so the more balanced the number of species and the lower the dominance among them, the greater the equability of this community. However, it should not be forgotten that this index is generated from the species table and does not verify the aggregation of individuals in the area.

A more diverse community with high equability indicates a more stable and balanced community. In the case of urban afforestation, this means greater diversity in urban vegetation and associated fauna, greater protection against possible pests and their proliferation, consequently less need for human intervention and greater biodiversity in a place where its construction process occurs with great losses of biodiversity.

With regard to the origin of the species found, the inventory showed equality between the species analyzed. Of the 130 species identified, 66 were exotic and another 64 were native, with 51 species native to the vegetation types in the study region and another 13 species native to other parts of Brazil. The literature consulted shows that it is common to find a greater number of exotic species than native ones in urban environments (ALBRECHT 1998; ANDREATTA et al., 2011; MENEGHETTI, 2003; MIRANDA; CARVALHO, 2009; PAIVA, 2009; SUCOMINE, 2010; TOSCAN et al., 2010) and even on other university *campuses* (LEAL; PEDROSA-MACEDO; BIONDI, 2009). In addition, there is little concern about the local phytophysiognomy for planting and few studies make this distinction, for example, Kurihara and Encinas (2003) found 35 species of Cerrado (phytophysiognomy of the study site), out of a total of 156 species surveyed on the *campus of* the University of Brasilia (UNB), similarly, Leal (2009) found 187 species, where 31 species were native to the local biome of his study, Finally, Lombardi (2003) found 187 species in his study, 88 of which were native to the region of Minas Gerais, identified on the *campus of* the Federal University of Minas Gerais, i.e. in no study do the local natives reach 50% and in other studies they are considered to be in the group of natives of Brazil.

Among the arguments justifying the priority of planting native species in urban afforestation are their resistance to pests, as they have already developed defenses to the region's pests and diseases, often reducing the need to use pesticides, their importance as food and shelter for native fauna and support for epiphytes that use them, allowing, in a way, the maintenance of coevolutionary processes between the different plant species, their pollinators, dispersers and their physical environment and the well-established relationship between the available nutrients and the nutrients needed by the trees, as the native species are already adapted to the local conditions. In addition to enhancing the value of native vegetation, planting native species reduces the risk of introducing invasive species that can cause problems for the conservation of biodiversity in a region.

It should be noted that some exotic species can cause ecological risks, as the introduction of species with invasion potential can cause biological impacts on forest remnants. Among the most abundant species found in this study (Table 1), *Ligustrum lucidum* was the second most abundant species. According to Matthews (2005), this species can become invasive and has already caused damage in several South American countries, including Brazil.

Another species is *Espathodea nilotica*, which had few individuals in this study, but which, according to the Horus Institute and I3N (2014), has also been causing serious environmental problems in forest remnants near urban areas. According to data from the Horus Institute and I3N (2014), the dispersal of this species, deliberately introduced for use in urban afforestation, occurs rapidly and is capable of competing with native plant species, preventing their regeneration. Dispersal by birds makes it impossible to prevent its spread. The institute's technicians suggest replacing the species with a non-invasive one.

According to the Global Invasive Species Database (GISD), the species *Terminalia catappa* L., *Archontophoenix cunninghamiana* H. Wendl. &. Drude, Pinus spp, Syzygium cumini, which were found in the inventory, are also among the invasive alien species in Brazil. According to the source, these species can cause damage to remnants of native vegetation and their dispersal by animals can make them easy to disperse and difficult to control. These species should also be avoided in urban areas and in some cases their replacement by other species should be studied.

Taking into account the choice of species native to phytophysiognomies found in the vicinity of the urban area, it would be advisable for the university to gradually implant cerrado trees, since it has a cerrado area on its territory. However, analyzing the local native species found in this sample of trees in the northern area of UFSCar, only 14 species (*Handroanthus ochraceus, Stryphnodendron adstringens, Qualea grandiflora, Anadenanthera falcata, Dalbergia miscolobium, Machaerium Acutifolium, Eriotheca gracilipes, Zanthoxylum roipholium, Cecropia pachystachya, Sterculia chicha, Hymenaea stigonocarpa, Baccharis dracunculifolia, Styrax ferrugineus, Myrcia bella*) are native to the cerrado. The presence of some native cerrado trees in the *campus* arboretum is due to their survival in the face of urbanization and not to their choice of planting.

Urban environments can be significantly altered, partly due to the local microclimate change inherent in a built environment and also due to the large-scale introduction of exotic plant species into parks and gardens (KONIJNENDIJK et al., 2005). Trees that are resistant to diseases and pests, adapted to the regional climate, adapted to the urban environment (tolerance to salt, photochemicals, stress, heavy metals, type of rooting, type of canopy, among others), as well as ornamental values and conservation of genetic resources are some of the parameters that should be considered when choosing tree species for urban areas and which are already used in some places in Europe (KONIJNENDIJK et al., 2005). One example is the afforestation program carried out in Montpellier, France, where the focus was on addressing both the environmental properties and functions maintained by tree biodiversity and human and social factors, including functionality, allergy reduction and others. This project began with a survey to identify wild species from that region in France that might have the potential to survive in urban areas. The research aimed to select species that would respond positively to the region's ecological factors (adaptation to climate, soil, etc.), interaction with society (toxicity, etc.), as well as ornamental values and reducing environmental risks (invasive plants) (KONIJNENDIJK et al., 2005).

The fruit-bearing species represent 42.1% of the individuals identified, distributed among 51 species (Table 2). Aroeira salsa makes up 18% of the fruit trees and is the most abundant. Among those present in the human diet, Jerivà is the most abundant, with 12.2% of the fruit trees, followed by 4.1% of guava trees and 3.9% of mango trees. Local native species account for 30% of the fruit trees and are divided into 16 species. Exotic species account for 46.1%, divided into 30 species. This data shows a preference for planting exotic fruit trees.

These species offer resources to fauna and help to attract and keep animals in the urban environment. They can attract various species of insects, birds and even small mammals (FRISCH AND FRISCH, 2005; KUHLMANN, 2012). With the use of native species, the attraction of native fauna can be greater and can form a more favorable environment for it.

Urban afforestation can play an important role in promoting and conserving the biodiversity that manages to survive in this highly disturbed environment (ALVEY, 2006). In this sense, promoting resources for fauna by planting high species richness can promote a great diversity of fauna and flora species in urban environments. Measures such as these to attract fauna are already provided for in the Sao Carlos afforestation plan and must be complied with (PMSC, 2009).

Table 2 Fruit species. Species (Scientific name of the species); Popular Name (Popular name of the species); NE (Origin of each species; 1: species native to the study region; 2: species native to Brazil; 3: exotic species from Brazil; 4: exotic and invasive species); N. Ind. (Number of individuals of each species).

Species	Popular Name	NE	N. Ind.
Schinus molle L.	Aroeira Salsa	2	194
Ligustrum lucidum W. T. Aiton	Tailor	4	156
Syagrus romanzoffiana (Cham.) Glassman	Jeriva	1	132
Nectandra megapotamica (Spreng.) Mez	Canelinha	1	55
Licania tomentosa (Benth)	Oiti	2	46
Psidium guajava L.	Guava tree	1	45
Mangifera indica L.	Hose	3	42
Eriobotrya japonica (Thunb.) Lindl	Loquat	3	38
Eugenia uniflora L.	Pitangueira	1	36
Michelia champaca L.	Magnolia	3	32
Dypsis lutescens (H. Wendl.) Beentje & J. Dransf.	Areca bamboo	3	28
Phoenix roebelenii O'Brien	Phoenix	3	27
Syzygium cumini (L.) Skeels	Jambolao	3	26
Morus nigra L.	Amoreira	3	23
Archontophoenix cunninghamiana H. Wendl. &. Drude	Seafortia	4	17
Roystonea oleracea L. H. Bailey	Imperial Palm	3	14
Persea americana Mill	Avocado tree	3	13
Araucaria angustifolia Raddi	Araucaria	1	13
Citrus limon L.	Limoeiro	3	13
Murraya paniculata (L.) Jack	False Myrtle	3	12

Table 2: Fruit species. *Continued.*

Species	Popular Name	NE	N. Ind.
Ficus benjamina L.	Ficus	3	11
Duranta repens L.	Drop of Gold	2	10
Malpighia glabra L.	Acerola	3	8
Schinus terebinthifolius Raddi.	Red mastic	1	7
Zanthoxylum roipholium Lam.	Clog	1	7
Dracaena marginata Lam.	Rainbow dracaena	3	6
Cecropia pachystachya Trecul	Embaùba	1	6
Eugenia pyriformis Camb.	Grapefruit	1	5
Annona squamosa L.	Anona	3	4
Anacardium occidentale L.	Cashew	1	4
Sterculia chicha A. St.-Hill. Ex Turpin	Meat	1	4
Genipa americana L.	Jenipapo	2	4
Dracaena fragrans (L.) Ker Gawl.	Waterwood	3	4
Coffea arabica L.	Coffee tree	3	3
Myrciaria jaboticaba (Vell.) O. Berg	Jabuticaba tree	1	3
Artocarpus heterophyllus Lam.	Jaqueira	3	3
Hymenaea courbaril var. stilbocarpa (Hayne) Y.T. Lee & Langenh	Jatobà	1	3
Muntingia calabura L.	Calabura	3	2
Codiaeum variegatum (L.) Rumph. Ex A. Juss	Croton	3	2
Chamaerops humilis L.	Mediterranean palm	3	2
Caryota urens L	Fishtail palm	3	2
Prunus persica (L.) Batsch	Peach tree	3	2
Punica granatum L.	Rome	3	2
Hymenaea stigonocarpa Mart. Ex Hayne var. stigonoc	J atobâ-do-cerrado	1	1
Styrax ferrugineus L.	Cerrado orange	1	1
Carica papaya L.	Papaya	3	1
Garcinia brasiliensis Mart.	Mangostao	2	1
Myrcia bella Cambess	Myrcia bella	1	1
Spondias purpurea L.	Seriguela	3	1
Terminalia catappa L.	Seven of Cups	3	1
Tamarindus indica L.	Tamarind	3	1
Total			**1074**

4.2 Field: Technical analysis of afforestation

At this stage, an analysis was made of each individual urban tree in relation to its size, conflicts with urban infrastructure, physical damage, need for management, the individual's health condition and planting location. Due to the large amount of information to be collected, the number of individuals inventoried was smaller than the number of individuals identified. Thus, out of the 3020 individuals inventoried, a technical analysis was carried out on 1422 individuals (Figure 11), which corresponds to 47% of the individuals inventoried, 36% of the urban tree cover in the study area and 24.2% of the urban tree cover in the whole of UFSCar - Sao Carlos. Table 3 shows the main results obtained, organized by species.

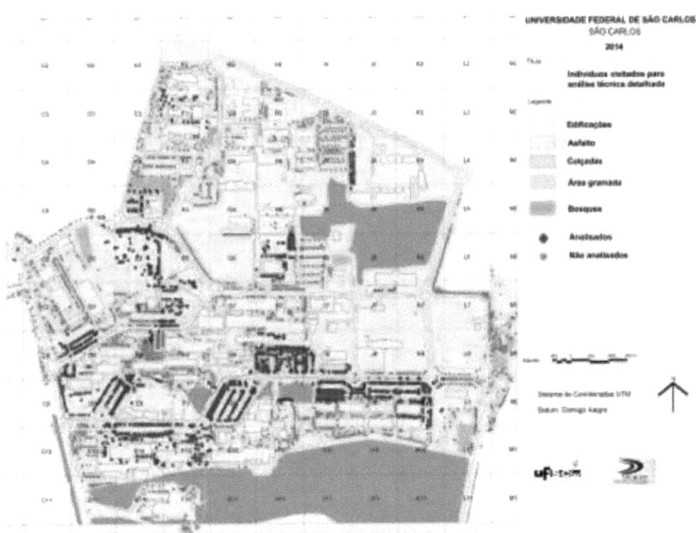

Figure 11: Individuals visited for detailed technical analysis during the inventory of urban afforestation in the North Area of UFSCar - São Carlos, SP, totaling 1422 individuals. Source: EDF UFSCar (2013) . Elaboration: Lessi, B. F. (2014).

Table 3 Data from the field survey of individuals organized by species. Species (scientific name of species); NE (Origin of each species; 1: species native to the study region; 2: species native to Brazil; 3: species exotic to Brazil; 4: exotic and invasive species); N. Ind. (Number of individuals of each species); N. Ind. (Number of individuals); CR (Root conflict with sidewalk); CE (Conflict with buildings); CP (Conflict with pedestrian walkways); CI (Conflict with lighting); PL (Light pruning), CFITO (Phytosanitary control); PFD (Physical damage due to pruning); T. CONF (Total number of individuals with each species); T. DEF (Total number of individuals with physical damage of each species).

Species	NE	N.Ind.	CR	EC	CP	CI	PL	CFITO	DFP	T.CONF	T.DEF
Caesalpinia pluviosa DC.	1	157	8	0	0	12	16	86	133	28	142
Ligustrum lucidum W. T. Aiton	4	126	3	2	0	23	14	47	85	35	88
Tabebuia heptaphylla (Vell.0 Toledo	1	107	4	9	1	3	10	7	79	22	79
Schinus molle L.	2	78	0	4	6	5	30	5	34	27	42
Tabebuia roseoalba (Ridl.) Sandwith	1	64	1	4	0	1	5	10	41	10	41
Syagrus romanzoffiana (Cham.) Glassman	1	49	0	1	0	0	2	0	2	2	2
Tibouchina granulosa (Desr.) Cogn	1	43	1	0	3	0	8	7	30	3	30
Nectandra megapotamica (Spreng.) Mez	1	41	3	0	0	13	10	11	31	17	31
Senna Siamea (Lam.) H.S. Irwin & R. C. Barneby	3	35	3	0	0	10	15	19	18	16	20
Handroanthus ochraceus (Cham.) Mattos	1	27	1	0	0	3	2	3	10	3	10
Cupressus lisitanica Mill	3	27	0	1	0	2	10	2	25	4	26
Michelia champaca L.	3	26	1	2	2	2	3	0	23	4	23
Psidium guajava L.	1	25	0	3	0	1	0	0	23	5	23
Anadenanthera pavonina L.	3	22	5	1	0	14	15	9	21	18	21
Syzygium cumini (L.) Skeels	3	21	0	0	0	0	0	17	21	0	21
Caesalpinia ferrea Mart.	1	17	5	0	0	2	1	0	9	2	9
Licania tomentosa (Benth)	2	16	0	0	0	1	4	0	12	1	12
Peltophorum dubium (Spreng.) Taub.	1	15	2	0	0	0	1	0	8	1	8

Keep going...

Table 3: Data from the field inspection of individuals organized by species. ***Continued.***

Species	NE	N.Ind.	CR	EC	CP	CI	PL	CFITO	DFP	T.CONF	T.DEF
Archontophoenix cunninghamiana H. Wendl. &. Drude	3	15	0	0	0	0	0	0	0	0	0
Phoenix roebelenii O'Brien	3	12	0	0	0	0	2	0	0	0	0
Dalbergia miscolobium Benth.	1	12	0	0	0	0	0	0	1	2	1
Callitropsis macrocarpa (Hartw. Ex Gordon) D.P.	3	11	0	0	0	0	0	0	0	0	0
Little *Caesalpinia pulcherrima* (L.) Sw	3	11	0	0	0	0	4	3	6	0	6
Tabebuia pentaphylla Hemsl.	3	11	1	0	1	1	1	0	4	2	4
Jacaranda brasiliana (Lam.) Pers.	1	10	0	0	0	1	1	0	4	1	4
Rhododendron simsii Planch	3	10	0	0	1	0	2	0	1	1	1
Stryphnodendron adstringens (Mart.) Coville	1	10	0	0	0	0	1	0	5	0	7
Eugenia uniflora L.	1	9	0	0	0	0	0	0	0	0	0
Guazuma crinita Mart.	1	9	0	0	0	1	1	0	1	1	1
Dypsis lutescens (H. Wendl.) Beentje & J. Dransf.	3	8	0	0	0	0	1	0	0	0	0
Machaerium Acutifolium Vogel	1	8	0	0	0	0	1	0	2	0	2
Albizia polycephala (Benth.) Killip	1	7	2	0	0	0	1	1	4	3	4
Citrus limon L.	3	7	0	2	0	0	0	0	3	2	3
Triplaris caracasana Cham.	3	7	0	0	0	0	1	0	0	0	0
Ficus benjamina L.	3	6	4	0	1	1	0	0	3	3	3
Delonix regia (Bojer ex Hook.) Ralf	3	6	4	0	0	0	0	0	5	0	5
Yucca elephantipes Regel.	3	6	0	0	0	0	1	0	0	0	0
Morus nigra L.	3	5	0	1	0	0	3	0	2	1	2
Anadenanthera falcata (Benth.) Speg	1	5	0	1	0	0	0	0	1	2	1
Cojoba sophorocarpa (Benth.) Britton & Rose	3	5	0	3	0	0	1	0	3	3	3

Keep going...

Table 3: Data from the field inspection of individuals organized by species. *Continued.*

Species	NE	N.Ind.	CR	EC	CP	CI	PL	CFITO	DFP	T.CONF	T.DEF
Schinus terebinthifolius Raddi.	1	5	1	0	0	0	0	1	3	2	3
Zanthoxylum roipholium Lam.	1	5	0	0	0	0	0	0	1	0	1
Mangifera indica L.	3	4	0	1	0	0	1	0	1	1	1
lagerstroemia indica L.	3	4	0	0	0	0	0	0	0	0	0
Persea americana Mill	3	4	0	1	0	0	0	1	4	1	4
Araucaria angustifolia Raddi	1	4	0	1	0	0	0	1	0	1	2
Malpighia glabra L.	3	4	0	0	0	0	0	0	1	0	1
Dracaena marginata Lam.	3	4	0	0	0	0	0	0	0	0	0
Senna pendula (Humb.& Bonpl. Ex Willd.)	1	4	0	0	0	0	3	0	0	0	3
Sterculia chicha A. St.-Hill. Ex Turpin	1	4	0	0	0	0	0	0	0	0	0
Dracaena fragrans (L.) Ker Gawl.	3	4	0	0	0	0	0	0	0	0	0
Eriobotrya japonica (Thunb.) Lindl	3	3	0	0	0	0	0	0	3	0	3
Plumeria rubra L.	3	3	0	0	0	0	0	0	3	0	3
Bauhinia variegata L.	3	3	0	0	0	0	0	0	2	0	2
Roystonea oleracea L. H. Bailey	3	3	1	0	0	0	0	0	0	0	0
Lafoensia glyptocarpa Koehne	2	3	0	0	0	1	1	0	3	1	3
Eriotheca gracilipes (K. Schum.) A. Robyns	1	3	0	0	0	0	0	0	0	1	1
Cassia grandis L.	2	3	0	0	0	0	1	0	2	0	2
Bougainvillea glabra Choisy	1	3	0	0	0	0	0	0	0	0	0
Furcraea foetida (L.) Haw.	1	2	0	0	0	0	0	0	0	0	0
Guazuma ulmifolia Lam.	1	2	1	0	0	0	0	0	0	0	0
Cedrela fissilis Vell	1	2	0	0	0	0	0	0	1	0	1
Coffea arabica L.	3	2	0	0	0	0	0	0	0	0	0
Hymenaea courbaril var. stilbocarpa (Hayne) Y.T. Lee & Langenh	1	2	0	0	0	0	1	0	0	0	0
Bauhinia L. sp	1	2	0	0	0	0	0	0	0	0	0
Codiaeum variegatum (L.) Rumph. Ex A. Juss	3	2	0	0	0	0	0	0	0	0	0

Keep going...

Table 3: Data from the field inspection of individuals organized by species. ***Continued.***

Species	NE	N.Ind.	CR	EC	CP	CI	PL	CFITO	DFP	T.CONF	T.DEF
Guarea guidonia (L.) Sleumer	1	2	0	0	0	0	0	0	0	0	0
Euphorbia leucocephala Lotsy *Bombax*	3	2	0	0	0	0	0	0	0	0	0
malabaricum DC.	3	2	0	0	0	0	0	0	0	0	0
Prunus persica (L.) Batsch	3	2	0	0	0	1	0	0	0	1	0
Enterolobium confortisiliquum (Vell.) Morong	1	1	0	0	0	1	0	1	1	1	1
Murraya paniculata (L.) Jack *Anadenanthera*	3	1	0	0	0	0	0	0	0	0	0
macrocarpa (Benth.) Brenan	1	1	0	0	0	0	0	0	1	0	1
Duranta repens L.	2	1	0	0	0	0	0	0	0	0	0
Pinus elliotti Engelm.	3	1	0	0	0	0	0	0	1	0	1
Eugenia pyriformis Camb.	1	1	0	0	0	0	0	0	0	0	0
Annona squamosa L.	3	1	0	0	0	0	0	0	1	0	1
Myrciaria jaboticaba (Vell.) O. Berg	1	1	0	0	0	0	0	0	1	0	1
Artocarpus heterophyllus Lam.	3	1	0	0	0	1	0	0	1	1	1
Ceiba speciosa (A. St.-Hill.) Ravenna	1	1	0	0	0	0	0	0	0	0	0
Baccharis dracunculifolia DC.	1	1	0	0	0	0	0	0	0	0	0
Muntingia calabura L.	3	1	0	0	0	0	0	0	0	0	0
Jacaranda macrantha Cham.	1	1	0	0	0	0	0	0	0	0	0
Chamaerops humilis L.	3	1	0	1	0	0	0	0	0	1	0
Punica granatum L.	3	1	0	1	0	0	0	0	0	1	0
Caliandra brevipes Benth.	2	1	0	0	0	0	0	0	0	0	0
Calliandra tweedii Benth.	2	1	0	0	0	0	0	0	0	0	0
Calistemon "imperiallis"	3	1	0	0	0	0	1	0	1	1	1
Jacaranda micrantha Cham.	2	1	0	0	0	0	0	0	0	0	0
Cassiafistula L.	3	1	0	0	0	0	0	0	1	0	1

Keep going...

Table 3: Data from the field inspection of individuals organized by species. ***Continued.***

Species	NE	N.Ind.	CR	EC	CP	CI	PL	CFITO	DFP	T.CONF	T.DEF
Hymenaea stigonocarpa Mart. Ex Hayne var. stigonoc	1	1	0	0	0	0	0	0	0	0	0
Styrax ferrugineus L.	1	1	0	0	0	0	0	0	0	0	0
Carica papaya L.	3	1	1	1	0	0	0	0	0	1	0
Tibouchina mutabilis (Vell.) Cogn	2	1	0	0	0	0	0	0	1	1	1
Garcinia brasiliensis Mart.	2	1	0	0	0	0	0	0	0	0	0
Swietenia macrophylla king	3	1	0	0	1	0	0	0	0	1	0
Myrcia bella Cambess	1	1	0	0	0	0	0	0	0	0	0
Calycophyllum spruceanum (Benth.) Hook. F. ex K. S	2	1	0	0	0	0	0	0	0	0	0
Qualea grandiflora	1	1	0	0	0	0	0	0	0	0	1
Araucaria calimnaris (J. R. Forst.) Hook	3	1	0	0	0	0	0	0	1	0	1
Sanchezia oblanga Ruiz & Pav	3	1	0	0	0	0	0	0	0	0	0
Spondias purpurea L.	3	1	0	0	1	0	1	0	0	1	0
Terminalia catappa L.	3	1	0	0	0	0	0	0	1	0	1
Melia uzedarach L.	3	1	0	0	0	0	0	0	0	0	0
Colubrina glandulosa Perkins	1	1	1	0	0	0	1	0	0	1	1
ID (Unidentified)	0	155	7	3	1	14	14	20	77	31	81
Dry/Dead	0	8	0	0	0	0	0	3	2	0	2
Start	0	46	0	0	0	0	0	0	0	0	0
Total		**1422**	**60**	**43**	**18**	**114**	**191**	**254**	**764**	**268**	**801**

4.2.1Size

Knowing the height (H) of trees and the maximum height that a species can reach is a very important factor for planning. The height of trees is an indispensable factor for planning, as it can interfere and conflict with various urban infrastructures, such as electrical wiring, public lighting and buildings. Each species has an average growth rate, which can be decisive when choosing a planting site.

The average height of the urban trees found was 6.39m. The maximum height was approximately 25m and the minimum was 0.1m (Figure 12).

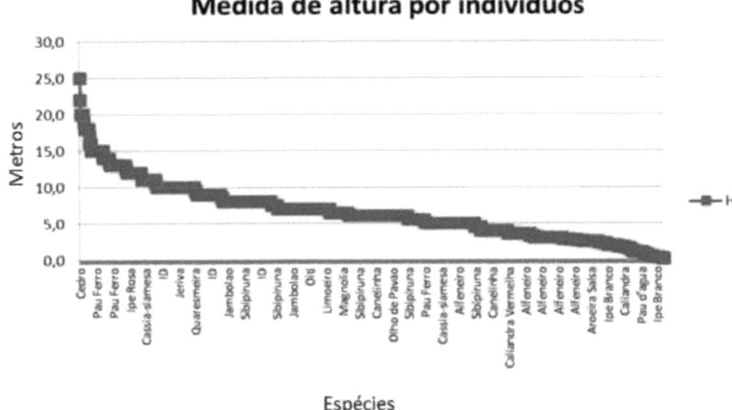

Figura 12: Height (H) of the individuals inventoried in the urban afforestation of the North Area of UFSCar - Sao Carlos, SP. Source: Lessi, B. F. (2014). Elaboration: Lessi, B. F. (2014).

The Sao Paulo Tree Planting Manual (PSP, 2005) considers small trees to be up to 5m tall, medium-sized trees to be between 5 and 10m tall and large trees to be over 10m tall. This study reveals that 41.8% of the trees are up to 5m tall, 39.9% are between 5 and 10m tall and 12.5% are over 10m tall. This shows that the urban forest is predominantly made up of small and medium-sized trees.

Figure 13 shows the distribution of the individuals inventoried according to height classification within the study area. Analyzing the image, it is possible to see that there is a greater concentration of small and medium-sized trees in the northern part of the study area and large trees in the southern part of the study area. This may be because the urbanization of this area occurred from the south to the north, so the northern areas are more recent and thus have had more recent planting, on the other hand, there may have been a choice of small species for the newly urbanized areas.

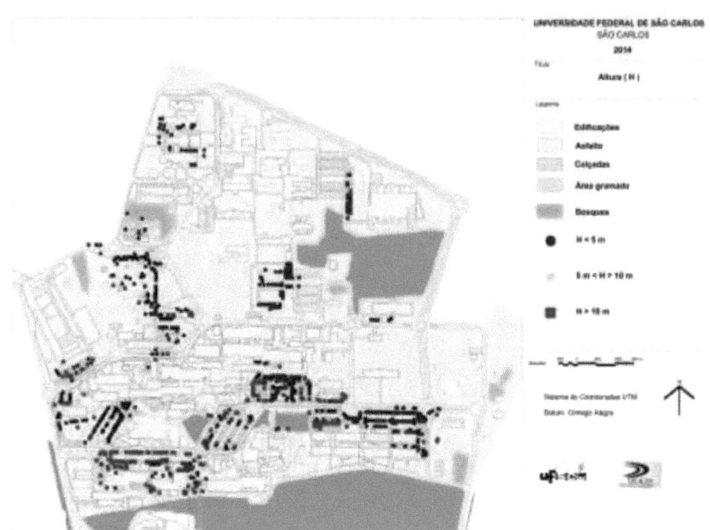

Figura 13: Individuals inventoried in the urban afforestation of UFSCar - Sao Carlos divided according to height categories (H) (class 1, in blue: H < 5m; class 2, in yellow: 5m < H > 10m; class 3, in red: H > 10m) for environmental compensation in Sao Carlos according to COMDEMA/SC Resolution No. 01/2012. Source: EDF UFSCar (2013), Manual de arborização urbana de Sao Paulo (2005) and Lessi, B. F. (2014). Elaboration: Lessi, B. F. (2014).

The average circumference at breast height was 0.65 m, with a maximum of 4 m and a minimum of 0.1 m. Figure 14 shows the PAC measurements. It shows that 84.8% of the individuals had a DBH of up to one meter and still many individuals below 0.5m (45.9%).

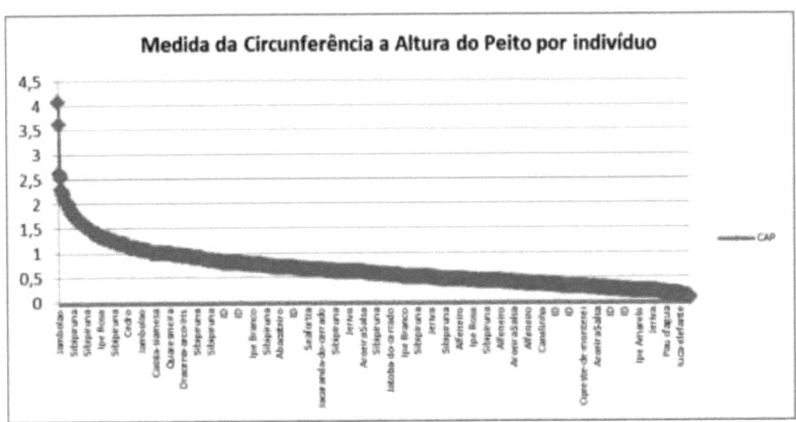

Figure 14: Circumference at breast height (CAP) of the individuals inventoried in the urban afforestation of the North Area of UFSCar - Sao Carlos SP. Source: Lessi, B. F. (2014). Elaboration: Lessi, B. F. (2014).

The compensation mechanism provided for by the Sao Carlos Municipal Council for the Defense of the Environment (CONDEMA 01/2012) is based on the origin of the species (native or exotic to Brazil) and its diameter at breast height (Table 4). With the data collected and the location of each tree (Figure 15), it is possible to predict and plan which trees will be cut down and how many will have to be planted to compensate for a possible felling. The data collected during the inventory was based on the CAP (Circumference at Breast Height), while CONDEMA is based on the DAP (Diameter at Breast Height) measurement. These measurements can be easily converted using the formulas: $C = 2.\pi.r$ and $D = 2.r$ (Where: C= Circumference perimeter; r = Circumference radius; D = Circumference diameter).

Table 4: Environmental compensation for tree felling in Sao Carlos, SP. DBH (Diameter at breast height)

Suppressed species	*Type of compensation*	*DBH < 0.15 m*	*0.15 m ≤ DBH ≤ 0.45 m*	*DBH > 0.45m*
Native	Planting and maintenance	03 seedlings	04 seedlings	08 seedlings
	To donate	10 seedlings	25 seedlings	60 seedlings
Exotic	Planting and maintenance	02 seedlings	04 seedlings	06 seedlings
	To donate	07 seedlings	20 seedlings	40 seedlings

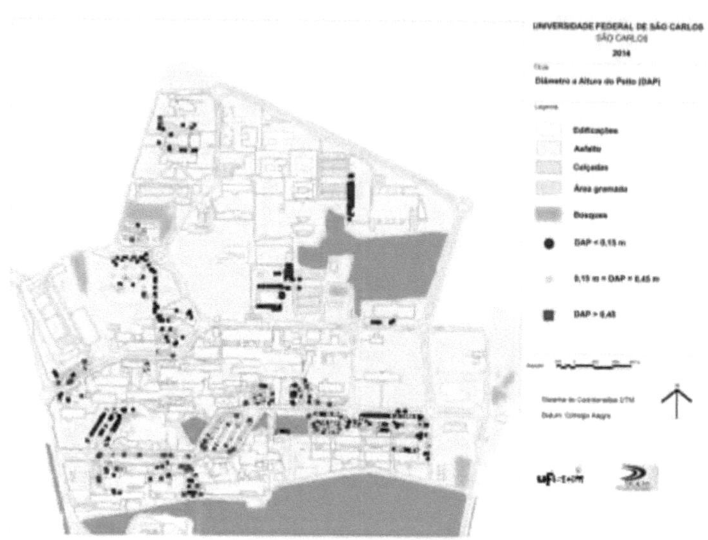

Figure 15: Individuals inventoried in the urban afforestation of UFSCar - Sao Carlos divided according to the categories of diameter at breast height (class 1, in blue: DBH < 0.15m; class 2, in yellow: 0.15m < DBH > 0.45m; class 3, in red: DBH > 0.45m) for environmental compensation in Sao Carlos according to COMDEMA/SC Resolution No. 01/2012. Source: EDF UFSCar (2013), COMDEMA/SC Resolution No. 01/2012 **and** Lessi, B. F. (2014). Elaboration: Lessi, B. F. (2014).

The image shows a predominance of individuals in the smaller size classes (DBH < 0.15m) and medium DBH (0.15m < DBH > 0.45m) in relation to the larger size class (DBH > 0.45m). It also shows a concentration of smaller individuals further north in the study area, which is probably due to new plantations, as these are more recently urbanized areas compared to the areas further south in the study area, where medium and large individuals prevail, or the choice of small species such as shrubs.

According to Aguirre Junior (2006) there has been a tendency in recent years to choose small species such as shrubs for urban tree planting because they adapt better to the urban environment and cause fewer conflicts, requiring lower management costs. However, larger species can offer greater environmental benefits and should not be replaced. The university has large areas that are home to large species and which should be used to achieve greater shade and better environmental quality. Angicos such as *Anadenanthera falcata,* for example, are a medium to large species found in the cerrado, which can be used well in the *campus* areas (LORENZI, 2008).

The larger the individual, the greater its height and consequently its circumference at breast height. An arboretum with large trees can bring more benefits to the urban environment, where all its functions will be maximized, such as transpiration, carbon sequestration and storage, shading, minimizing pollution and influencing the urban climate. On the other hand, when trees are extremely large, they become difficult to manage, such as removing dry branches and pruning, and there is also a greater risk of damage in the event of a fall.

The measurement of the first bifurcation (HB) of the individuals showed an average of 1.4m, a maximum of 11m and a minimum of zero (Figure 16). The data shows that most of the individuals have a HB of less than 2m (81.6%) and 60% of these have a CAP of up to 0.5m. Many individuals had branches coming from the ground, i.e. with bifurcations at ground level, which were considered to be zero; these individuals amount to approximately 13%. According to the Erechim Master Plan for Afforestation (PME, 2011), individuals with a HB of less than 1m are the result of poor management of seedlings, inadequate pruning or are young or small individuals. In addition, individuals with low HB can cause problems for urban infrastructure, such as conflicts with pedestrians, lighting and traffic, among others.

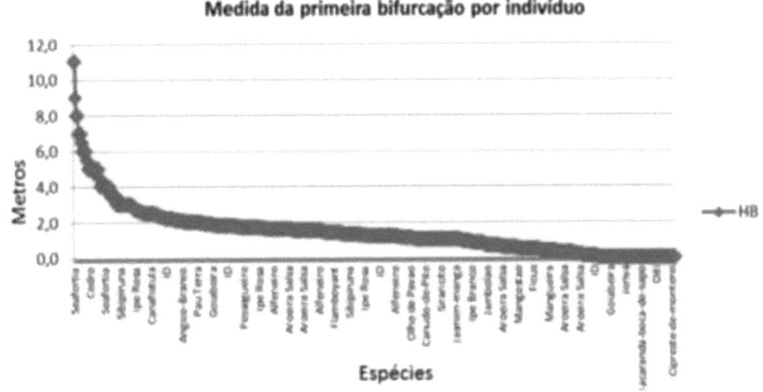

Figure 16: Height of the first bifurcation (HB) per individual. of the individuals inventoried in the urban afforestation of the North Area of UFSCar - Sao Carlos SP. Source: Lessi, B. F. (2014). Elaboration: Lessi, B. F. (2014).

Some authors suggest that seedlings should have a minimum size so that they have a greater chance of survival and thus less need for management, avoiding future conflicts (CEMIG, 2011; MATOS; QUEIROZ, 2009; PME, 2011; PSP, 2005). The Erechim Urban Afforestation Master Plan (PME, 2011) also suggests that saplings should be at least 2m high and 1.5m from the first fork. According to the Companhia Energética de Minas Gerais (CEMIG, 2011), seedlings should be at least 2.5 m in height at the first fork, and 0.05 m in diameter at breast height (CAP = 0.15). The São Paulo City Council's Urban Afforestation Manual (PSP, 2005) suggests that seedlings should be 2.5 m high, with a height of the first fork of 1.8 m and a diameter at breast height of 0.03 m (DBH = 0.1 m). The Sao Carlos Tree Planting Manual (PMSC, 2009) specifies seedlings with a minimum height of 2.0m, 1.5m from the first fork and a diameter at breast height of 0.02m.

The sample also reveals individuals planted with measurements below those suggested in the literature, with 9.8% below the indicated height of 2 meters, 72.5% of individuals with the first bifurcation below 1.5m and in relation to CAP, all have the minimum suggested. It also reveals newly planted seedlings with heights of just 0.1 centimeters, with 6% having heights of up to 1 m, which would be half of what is indicated. The planting of very young individuals, together with the stresses encountered in the urban environment, can greatly increase plant mortality, which is why they are not recommended.

4.2.2The relationship between afforestation and urban infrastructure

The inventory revealed 269 individual trees conflicting with urban infrastructures (Table 3; Figure 17). We found 60 conflicts with the sidewalk, 43 conflicts with buildings, 2 conflicts with electrical wiring, 114 with street lighting and 18 with pedestrian sidewalks. The species with the most conflicts were privet (*Ligustrum lucidum*) conflicting with lighting, flamboyant (*Delonix regia*) conflicting with the sidewalk and pink Ipê (*Tabebuia heptaphylla*) conflicting with buildings.

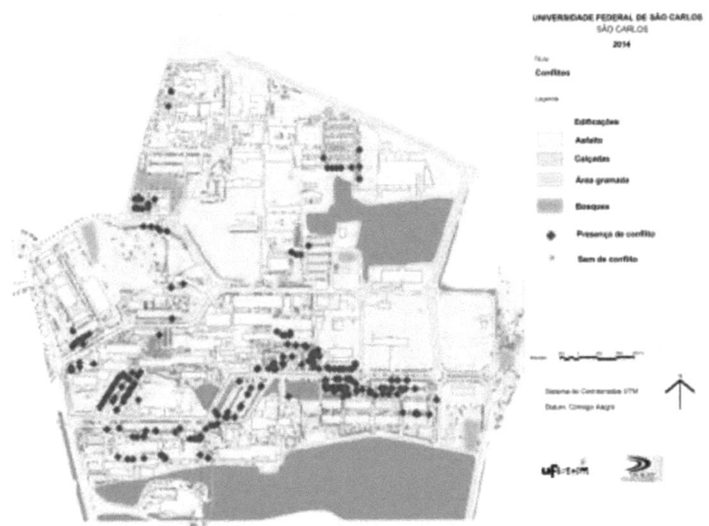

Figure 17: Individuals in the inventory of urban trees that presented some kind of conflict with the urban infrastructure of the North Area of UFSCar - Sao Carlos. Source: EDF UFSCar (2013) . Elaboration: Lessi, B. F. (2014).

Conflicting individuals account for 18.9% of the total tree cover. Compared to other studies, which found percentages of 25% (MIRANDA; CARVALHO, 2009), 35.1% (MENEGHETTI, 2003), 42% (SCHUCH, 2006) and 45% (ROCHA; LELES; NETO, 2004), the results of this study show a relatively low percentage of conflicts.

In relation to conflict with electrical wiring, for example, the sampling carried out shows 2 individuals in conflict (Figure 18), while other studies that have carried out this type of assessment have shown situations such as 18.7% in conflict with wiring, surveyed by Mazioli, 2012, 41% surveyed by Pires et al., 2010, 50.2% presented by (SILVA et al., 2002) and even 61% obtained by Toscan et al. (2010).

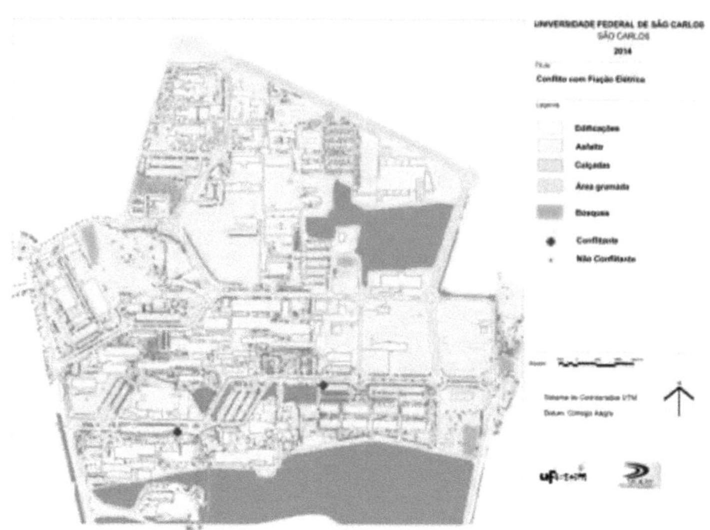

Figure 18: Individuals in the urban forest inventory who presented some conflict with the electricity transmission network in the North Area of UFSCar - Sao Carlos. Source: EDF UFSCar (2013) . Elaboration: Lessi, B. F. (2014).

The results showed a total of 3% of individual trees in conflict with buildings (Figure 19). This figure is close to that found in other studies, 2% (PIRES et al., 2010) and 6% (SCHUCH, 2006), which shows concern about these conflicts, since they can cause great expense if the tree falls on the building, in addition to the problem of maintaining gutters and roofs. According to Guzzo and Carneiro (2008), if the mains connections are conventional, it is advisable to plant small trees below the mains, but in the case of protected and isolated connections, medium and large species can be planted and the minimum distance from the pole would be 4m.

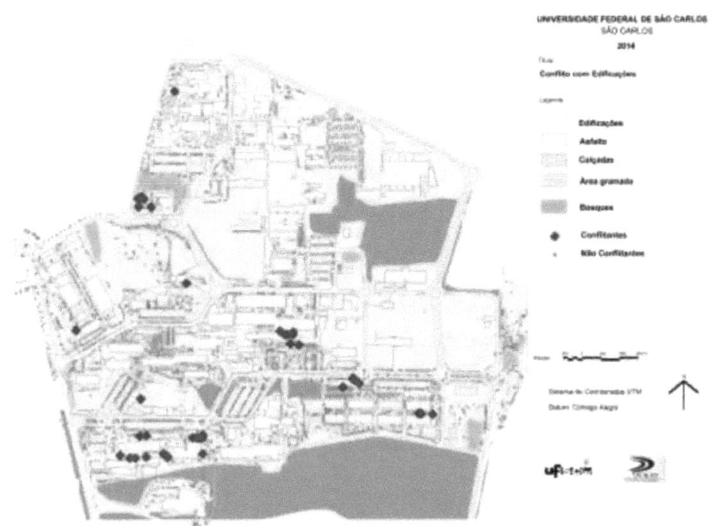

Figure 19: Inventoried individuals in the urban forest that presented some conflict with buildings in the North Area of UFSCar - Sâo Carlos. Source: EDF UFSCar (2013) . Elaboration: Lessi, B. F. (2014).

Few individuals showed conflicts with the sidewalks (3%) (Figure 20), in comparison with other studies which found 30.4% (BENATTI et al., 2012) and 37% (TOSCAN et al., 2010), but which were carried out in urban centers. The presence of individuals far from the sidewalk in large lawns also reflects the low percentage of conflicts with the sidewalk compared to cities.

Choosing the right species for planting can also have an influence, for example, by choosing ipês, which are trees with roots that grow less than *Delonix regia* (Flamboyant) and *Ficus benjamina* (Fig tree), which are species considered unsuitable for urban tree planting and which have already been found damaging sidewalks (PIRES et al., 2010).

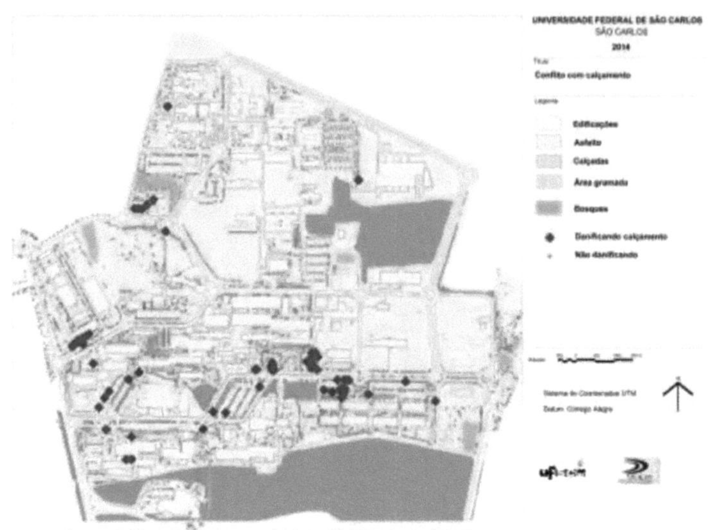

Figure 20: Inventoried urban tree species that presented some conflict with the sidewalk in the North Area of UFSCar - Sao Carlos. Source: EDF UFSCar (2013) . Elaboration: Lessi, B. F. (2014).

Around 8% of the individuals surveyed had conflicts with the public lighting on *campus* (Figure 21). This frequency is close to the 6% found in the literature (PIRES et al., 2010). The proximity between lampposts and trees and the average height of 6.39m (found in this study), very close to the height of the *campus lampposts*, may favor the presence of this conflict, since the lampposts are approximately 5m high. The privet (*Ligustrum lucidum*) has the highest number of conflicts with lighting due to its planting very close to the lampposts and its development, which allows the crown to grow at the height of the lamppost, obstructing the lighting.

Planting at a minimum distance of 4m from the lamppost (GUZZO; CARNEIRO, 2008) and choosing species with average growth above the can reduce the incidence of this conflict, since the lampposts should be below the treetops (PME, 2011), as can pruning very close individuals with crown growth at the height of the lighting.

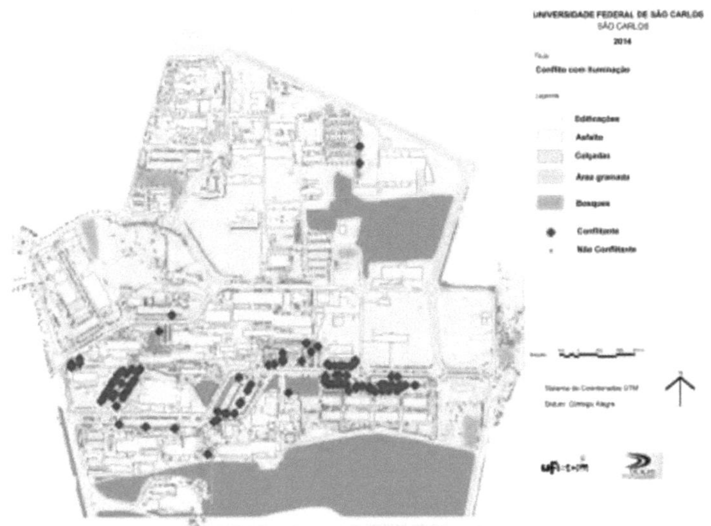

Figure 21: Inventoried urban tree species that presented some conflict with the lighting in the North Area of UFSCar - Sao Carlos. Source: EDF UFSCar (2013) . Elaboration: Lessi, B. F. (2014).

The presence of conflicts with the pedestrian walkway was low, with only 18 conflicts (Figure 22). The presence of this conflict may be directly linked to the presence of small individuals (up to 5 meters) and individuals with the first fork lower than indicated. The proximity of individuals with these characteristics to the sidewalk can increase the occurrence of this conflict. This conflict can force pedestrians to alter their path, often having to walk across the street, which can pose risks to their physical integrity. It also deprives pedestrians of benefits such as shade. Another important issue is the difficulty this can cause for visually impaired people. In this scenario, pruning these individuals, in order to resolve the conflicts, would be the most appropriate course of action.

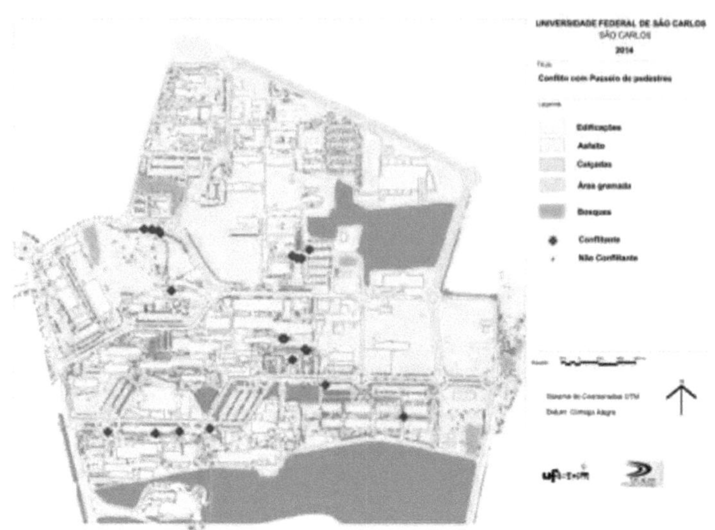

Figure 22: Inventoried urban trees that presented some kind of conflict with the pedestrian walkway in the North Area of UFSCar - Sâo Carlos. Source: EDF UFSCar (2013) . Elaboration: Lessi, B. F. (2014).

Although the number of conflicts is relatively small, this is not the case with records of physical damage caused by poorly executed pruning. Of the trees inventoried in UFSCar's North Area, 53.7% had physical damage (Figure 23). Other studies also show high percentages of physical damage caused by pruning. Pires et al. (2010) observed 64% of the trees in Goiandira (GO), Schuch (2006) with 39% observed in Sao Pedro do Sul (RS), Toscan et al. (2010) with 33% observed in Foz do Iguaçu (PR) and Milano (1988) who observed 28.8% of the trees with damage caused by pruning in Maringà (PR).

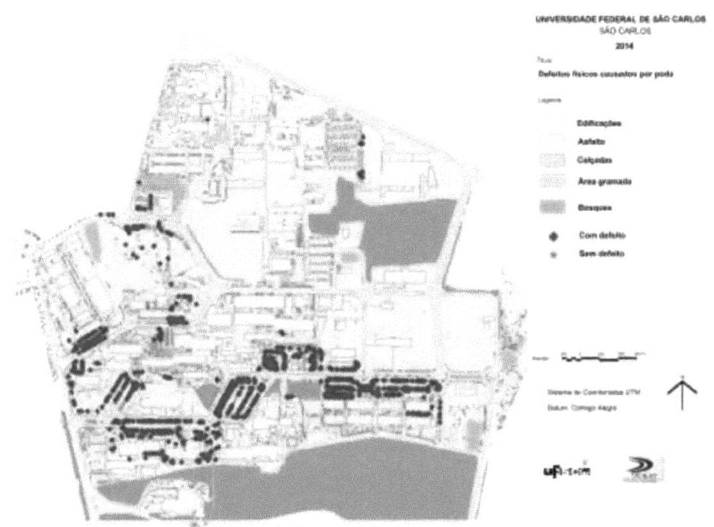

Figure 23: Individuals inventoried that showed physical damage caused by pruning in the urban afforestation of the North Area of UFSCar Sao Carlos. Source: EDF UFSCar (2013) . Elaboration: Lessi, B. F. (2014).

The occurrence of this physical damage reveals the need to improve tree management practices, as pruning must be carried out with extreme care so as not to cause problems for the trees. In addition to improving pruning techniques, it will be necessary in the future to adopt practical criteria for choosing the right species for different locations, depending on the type of limitation, such as lower-growing species under power lines and lighting points, with smaller canopies near buildings.

Conflicts between urban forestry and urban infrastructure should always be avoided, but it is also necessary to maintain a well-wooded urbanization. To this end, there are several manuals and books on how to avoid conflicts (MASCARÓ; MASCARÓ 2010; MATOS; QUEIROZ, 2009). In addition, the planning and construction of infrastructure, leaving space for trees and the choice of the right species can be determining factors in the emergence of conflicts.

The low incidence of conflicts may in some way be related to the large number of trees with physical damage caused by pruning. Bearing in mind that pruning is not a natural process, but man-made to resolve conflicts or for landscaping purposes, some conflicts may have already been resolved by this practice. Thus, the lack of planning can have a direct impact on this data, since the location of the planting, the correct choice of species and the construction of an infrastructure that takes tree planting into account are all practices that should be the subject of urban planning. A master plan for urban tree planting can take several of these aspects into account (PME, 2011), which can help *campus* managers make decisions.

4.2.3Need for pruning

With regard to the need for pruning, the inventory revealed that 13.2% of the trees currently require light pruning (Figure 24). This percentage is lower than that found by other authors, such as Melo and Severo (2007) who found 29.27%, Benatti et al. (2012) who found 33%, Faria; Monteiro; Fisch, 2007 who found 45% and Milano (1988) who observed 51.5% of the trees in Maringà in need of light pruning in his study.

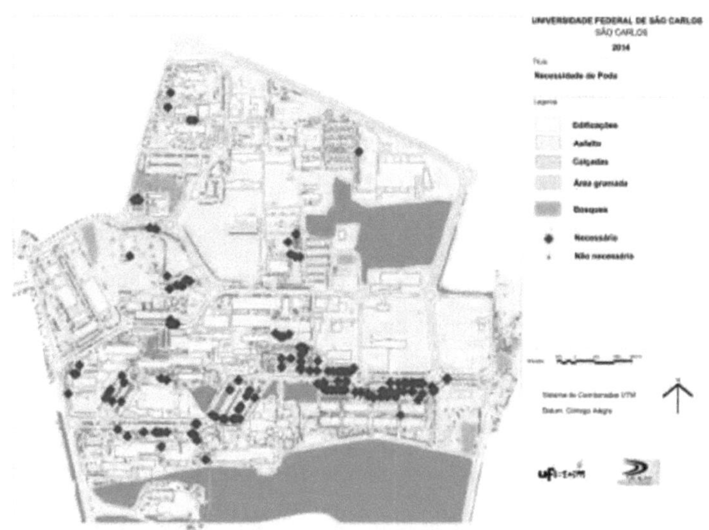

Figure 24: Individuals in need of pruning in the inventory of urban trees in the North Area of UFSCar - São Carlos. Source: EDF UFSCar (2013) . Elaboration: Lessi, B. F. (2014).

Pruning is an operation carried out to adapt the tree to the conditions of the urban environment. Various pruning techniques can be used (MASCARÓ E MASCARÓ, 2010), such as corrective pruning, with the aim of making the tree more upright, with no lateral sprouting or to remove dry branches in order to prevent falls, conduction pruning to better adapt the tree to the urban environment, with the removal of branches, resolving conflicts, cleaning or maintenance pruning to eliminate dry and diseased branches, which can be more drastic and heavy pruning, in order to remove parts of the trees that are putting people or property at risk. In addition to different pruning techniques, there are techniques for correctly cutting the branches in order to achieve better healing of the cut and the correct times for each group of species with different behavior and life cycles (PSP, 2002). The correct time for pruning is also related to nesting birds. Law 9605/98, art. 29 x 1, items I and II, on environmental crimes, provides for the protection of nesting birds by prohibiting the removal of their nests during this period, including by pruning (PSP, 2002).

4.2.4Positioning and permeable planting area

The analysis of the distance from the curb was only carried out on the individuals around the library street (660 individuals) (Figure 25). Due to the large number of individuals in grassy spaces, and distant from the armoring and sidewalk, many individuals were disregarded for this measure, where distances were considered up to 20m (Figure 26).

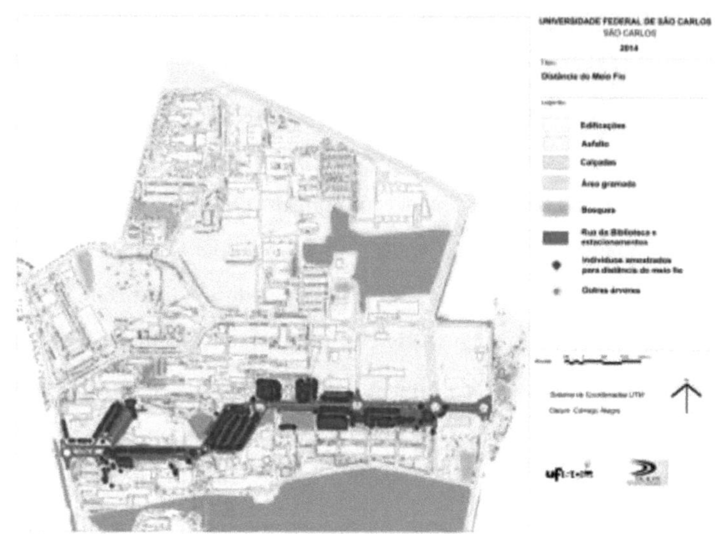

Figure 25: Individuals analyzed for distance from the curb in the urban tree inventory of the North Area of UFSCar - Sâo Carlos. Source: EDF UFSCar (2013) . Elaboration: Lessi, B. F. (2014).

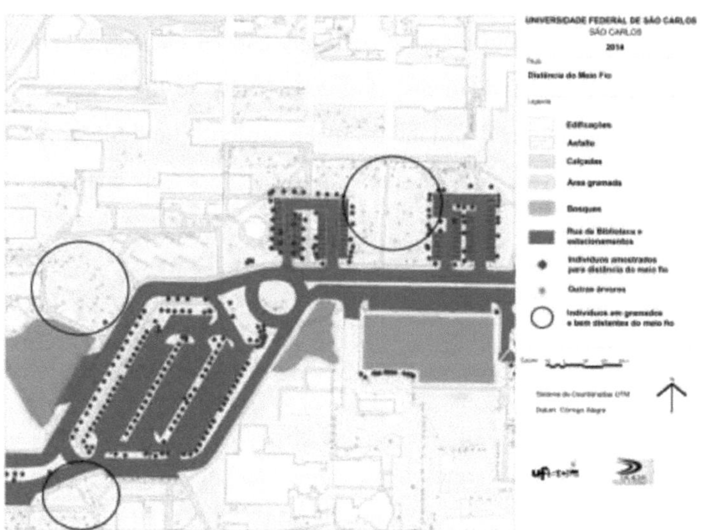

Figure 26: Individual urban trees far from the street, located on lawns in the North Area of UFSCar - Sao Carlos. To the northeast of the image is the Civil Engineering Department, and to the southwest is the North Area Amphitheater - Bento Prado. Source: EDF UFSCar (2013) . Elaboration: Lessi, B. F. (2014).

The average found was 1.95m with a minimum of 0.1m, a great distance if compared with the work of Albrecht (1998) who found averages of 0.21m in Sao Carlos. This can be considered a safe distance to avoid damage to the streets due to root growth and to avoid conflict with vehicles, since other studies indicate shorter distances, such as Albrecht (1998) and the Sao Carlos Urban Tree Plan (PMSC, 2009), which suggests a minimum of 0.5m, and the Sao Paulo Urban Tree Technician's Manual (PSP, 2005), which suggests a minimum of 0.3m.

Due to the existence of water stress and soil compaction in the urban environment (URBAN, 1989), the existence of a free area, i.e. a permeable area at the planting site of each individual surveyed, was assessed. The sample revealed that 86% of the trees had the recommended clearance area or above (Code 1), 9% had an average permeable area (Code 2), 3% had a clearance area below the recommended (Code 3) and only 1% of the trees had no permeable area. The large number of individuals with Code 1 clearings may have been due to the large number of trees on lawns (Figure 17). Individuals in codes 3 and 4 should have their open areas adapted to improve their environmental condition (Figure 27).

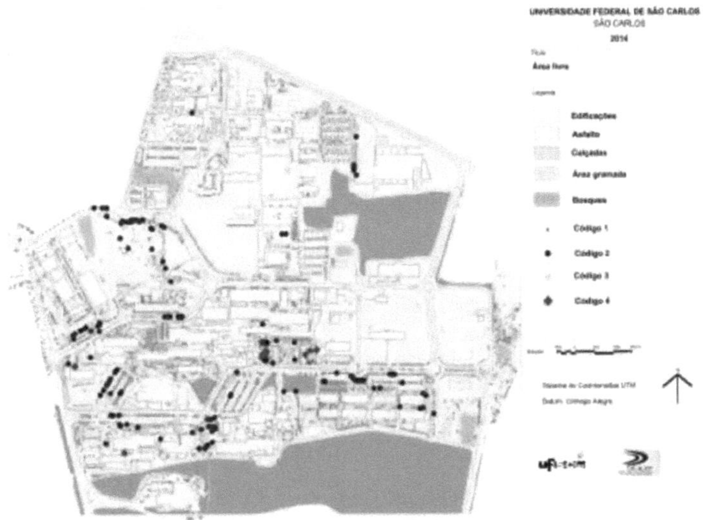

Figure 27: Free and permeable area below each individual in the inventory of urban trees in the North Area of UFSCar - São Carlos. Source: EDF UFSCar (2013) . Elaboration: Lessi, B. F. (2014).

With the health of the plants in mind, in order to avoid possible deaths, falls and risks on *campus*, trees with low or no permeable areas around their base (base of the trunk) should undergo intervention and, as far as possible, the opening and increase of permeable areas should be evaluated in the near future.

The positioning of the planting in relation to the buildings is also very important to avoid conflicts. When individuals are in close proximity to buildings, they should be in good health so that the risk of falling is as low as possible, avoiding material damage.

As with the measurements of distance from the curb, the distances from the buildings were measured up to 20m from the buildings, where individuals with distances over 20m from the buildings were considered to be over 20m. Individuals that were too isolated or far from any building were disregarded. The data shows an average of 17.6m where 77% of the measurements are 20m or more, i.e. most of the trees are far from buildings.

Trees near buildings are important for maintaining the temperature, which can save energy, especially in hot weather. Another point would be to protect the building from wind and noise.

This distance can bring safety on the one hand, as few trees reach a height of 20 m or more and can therefore bring some risk in the event of a fall. On the other hand, fewer trees nearby can mean less shading for buildings, which can cause thermal discomfort during periods of heat and drought due to higher temperatures.

4.2.5General state of afforestation

With the presence of various stress factors in the urban environment (URBAN, 1989), the vegetation in this environment can become less resistant to pests and diseases, and also, with the large occurrence of groups of trees of the same species, pests and diseases can spread easily (SANTAMOUR, 1990). It was found that 20% of the individuals had different phytosanitary control needs (Figure 28).

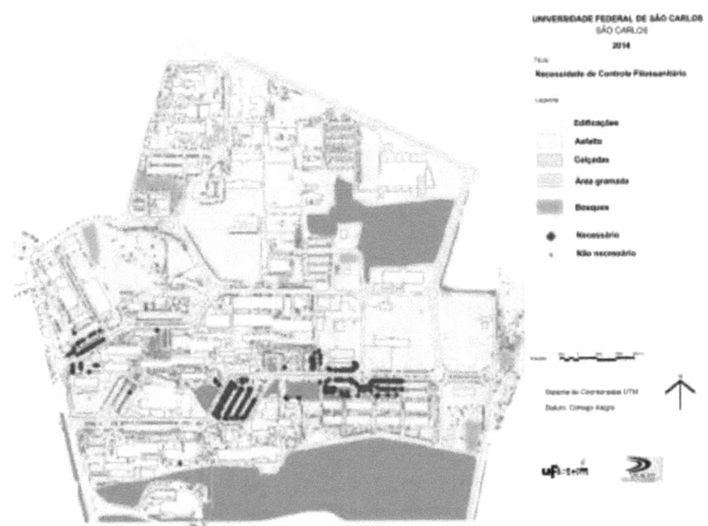

Figure 28: Individuals in need of phytosanitary control recorded in the inventory of urban trees in the North Area of UFSCar - Sao Carlos. Source: EDF UFSCar (2013) . Elaboration: Lessi, B. F. (2014).

Plant health can be directly related to the presence of pruning on trees (MARTINS; ANDRADE; ANGELIS, 2010). The data collected shows that 80% of individuals in need of phytosanitary control have physical damage caused by poorly executed pruning. According to Albrecht (1998), some damage begins with poorly executed and poorly healed pruning that leaves the inside of the plant exposed, which can become a source of entry and development of pests and diseases (Figure 29).

Figure 29: Physical damage and poorly executed pruning can contribute to the installation of pests and diseases.

diseases. Individual located in the parking lot in front of the Area Norte - Bento Prado Amphitheatre at UFSCar - Sao Carlos. Photo: Lessi, BF (2013).

The species with the most problems of this kind was Sibipiruna, with 31% of the individuals requiring control. This fact has also been verified by other authors (ALBRECHT, 1998; ALBERTIN et al., 2011; DUARTE et al., 2008), corroborating that pruning can lead to the installation of pests and diseases on plants.

In many cases, the need for phytosanitary control was only assessed by the presence of wounds, hollow holes in the trunks, the presence of dry branches, for individuals that were not healthy, among others (Figure 30). Therefore, the identification and determination of the treatment should be done by a specialist.

Figure 30: Wounds, the presence of concavities and poorly executed pruning can contribute to the installation of pests and diseases. A - Individual located in the parking lot in front of the Area Norte Amphitheatre - Bento Prado; B - Individual located in front of Classroom Building 5. Photo: Lessi, BF (2013).

Only termites and ants were found visibly (Figure 31). Termites affected 11% of the individuals in need of control, while ants affected 1.4% of the individuals observed. No species preference was found for ants. Of those with termites, 39.3% were Ipê-rosa and Sibipiruna. Other studies have also found Sibipiruna to be the species most affected by termites in urban areas (ALBERTIN et al., 2011; DUARTE et al., 2008), which reveals the need for special care when choosing this species for planting.

Figure 31: Visible occurrence of pests. Individual located in the parking lot of the Center for Biological and

Health Sciences (CCBS) in the North Area of UFSCar - Sao Carlos. Photo: Lessi, B. F. (2013).

The proposed classification showed that 75% of the trees sampled in the area north of UFSCar are in good condition and have good integrity. Due to urban stresses, the average lifespan of a tree in an urban environment varies from 10 to 25 years, so keeping them in good integrity means maintaining good environmental conditions so that they survive as long as possible (URBAN 1989).

Even though most of the trees are in good condition (Figure 32), some of them have some negative aspects that could, over time, affect their physical integrity. For example, 3.5% of these trees have a lower clearance than indicated, 10.5% need phytosanitary control and 50% have physical damage caused by pruning. With these negative aspects, almost half of the trees have a tendency to get sick, which must be taken into account for good management.

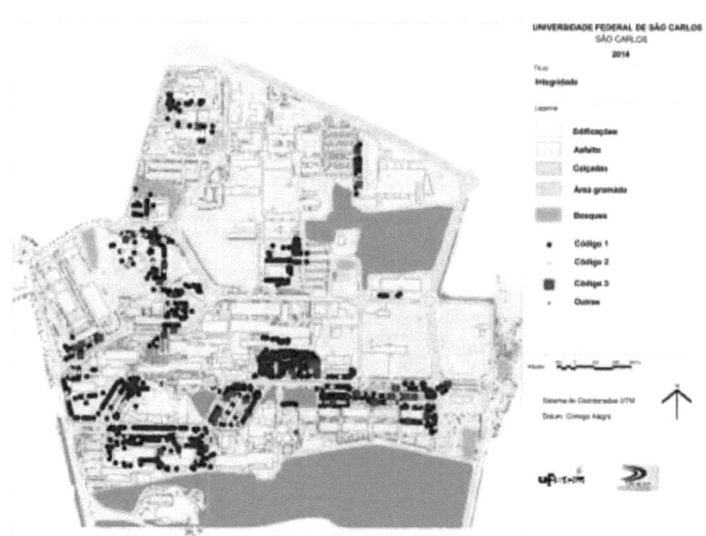

Figure 32: General condition of the individuals inventoried in the urban afforestation of the North Area of UFSCar - Sao Carlos. Code 1 (good, blue); code 2 (intermediate, yellow); code 3 (bad, red); Others (trees not inventoried, green). Code 1 (good, blue); code 2 (intermediate, yellow); code 3 (bad, red); Others (non-inventoried trees, green). Source: EDF UFSCar (2013) . Elaboration: Lessi, B. F. (2014).

With regard to individuals with low integrity, which amount to 3.8%, these are few individuals, but they require a great deal of attention. Among these individuals with low integrity, 63% need physical control, 72% have physical damage from pruning and 61% are at risk of codes 3 and 4, which are the highest. This data shows the great influence that diseases, pests and pruning can have on the integrity of a plant (MARTINS et al., 2010). Eight of these individuals were dry and apparently dead, so the need to remove them and replace them with a new individual should be analyzed, so that they don't cause any damage to the university in the event of a fall.

In order to highlight these risks, a "Risk Analysis" category was created, where each individual was given a code according to their risk of a possible fall and the damage they could cause by falling. This assessment is extremely important for decision-making in tree management, in order to avoid material damage to the university and risks to people's health. Using this information, it is possible to prioritize the individuals with the greatest need for attention and the most urgent intervention. In the case of dry and dead trees, they were given the highest risk category (4), indicating the need to intervene in these individuals, as they are more likely to fall (Figure 33).

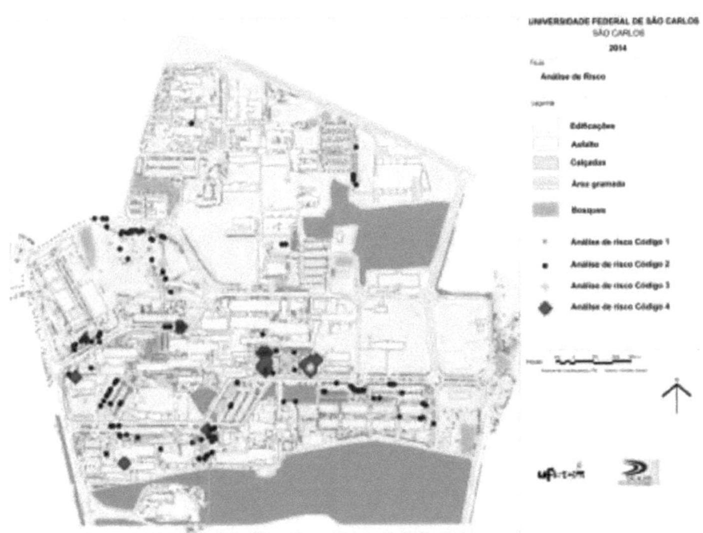

Figure 33: Risk analysis of urban trees inventoried in the North Area of UFSCar - Sâo Carlos . Source:

EDF UFSCar (2013) . Elaboration: Lessi, B. F. (2014).

Trees with a risk code between 3 and 4, which are the most at risk, represent 3.6% of the individuals inventoried. Of these, 50% need physical control and 66.6% have physical damage caused by pruning.

Thinking about all the benefits that trees can bring to an urban environment (temperature maintenance, microclimate, pollutant retention, carbon dioxide consumption, shading, etc.) and how important each individual tree can be to this environment due to the location in which it is found and the use of this location by more or fewer people, the degree of functionality of each tree was created. It was considered that each tree has its minimum environmental functionality, so all trees, without exception, have their importance and together can maximize their functions.

The survey shows that 30% have a level of functionality of code 1, another 51% with a level of code 2 and finally 17% with the maximum level of functionality (code 3) (Figure 34).

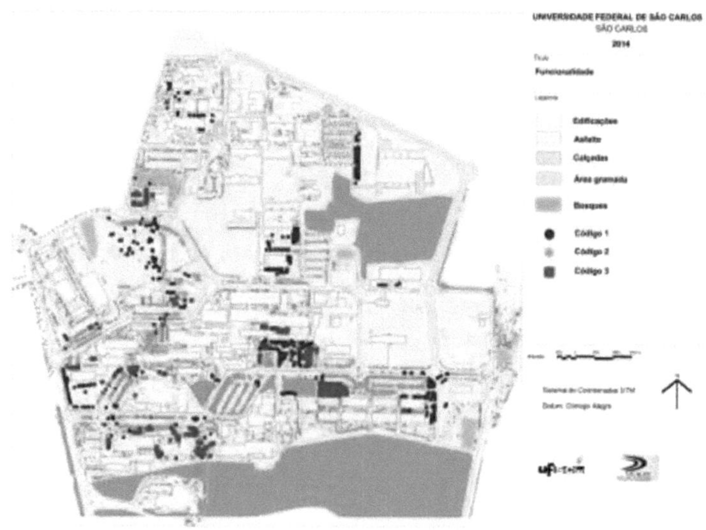

Figure 34: Degree of functionality of the individuals inventoried in the urban afforestation of the North Area of UFSCar - Sao Carlos. Source: EDF UFSCar (2013) . Elaboration: Lessi, B. F. (2014).

Some of the trees are in grassy areas, away from pedestrians and cars, which explains the 30% with functionality code 1. It is important to note that this does not mean that these trees are unimportant, as there is no doubt that they help to mitigate climatic extremes and the local hydrological cycle (evapotranspiration, infiltration). The presence of mainly well-wooded parking lots, together with the large circulation of cars and people, can explain the large number of trees with functionality code 2 (51%). The low number of individuals with maximum functionality can be explained by the low number of tree-lined sidewalks and trees near the buildings (Figure 35), where a large number of people circulate. On the other hand, these individuals with maximum functionality must be maintained with high physical integrity, as well as being monitored and managed regularly, since they have a greater chance of causing more serious damage.

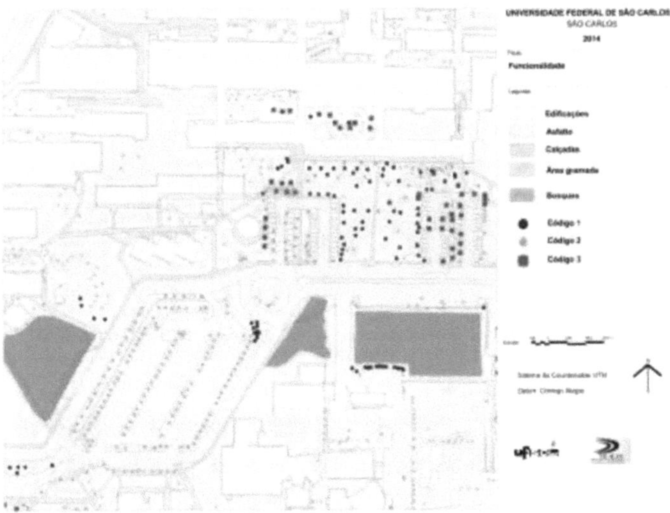

Figure 35: Degree of functionality of the individuals inventoried in the urban afforestation of the North Area of UFSCar - Sâo Carlos, in detail (Parking lot in front of the North Area Amphitheater - Bento Prado and further north the Civil Engineering Department). Source: EDF UFSCar (2013) . Elaboration: Lessi, B. F. (2014).

4.3 Land use and occupation and the urban vegetation canopy

The good coverage provided by the urban vegetation canopy is very important for mitigating the heat in a university environment, where many people walk or cycle.

The area studied within the *campus* is approximately 817,000 m^2 (81.7 ha). Within this area, in addition to the tree survey, land use and occupation mapping was carried out (Figure 36 and Table 5).

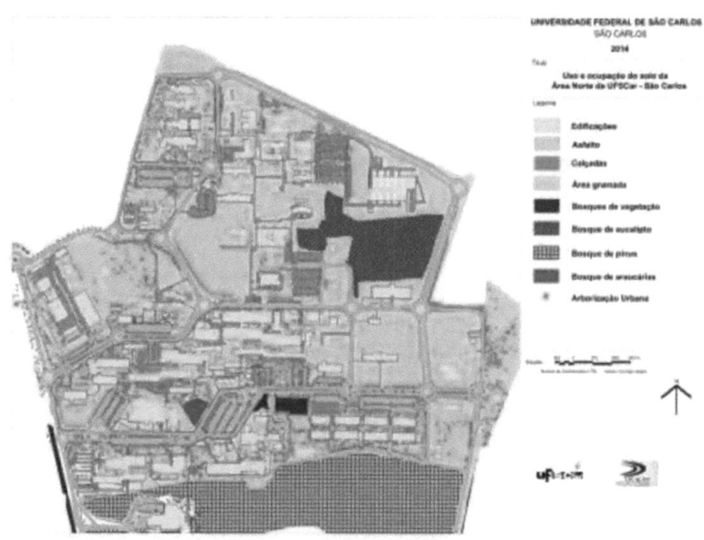

Figure 36: Land use and occupation in the North Area of UFSCar - Sao Carlos. Source: EDF UFSCar (2013) .

Elaboration: Lessi, B. F. (2014).

Table 5: Land use and occupation of the urban vegetation canopy in the North Area of UFSCar - Sao Carlos.

Land use and occupation	Area (m^2)	%
Area studied	817.000,00	
Built-up area	95.551,04	11,7
Roads (Asphalt)	109.118,83	13,4
Sidewalks	78.696,95	9,6
Lawn areas	370.125,67	45,3

Woods	128.187,43	15,7
Canopy	244.342,49	29,9

The built-up area represents 11.6% of the area, the paved area 13.3% and the sidewalk area only 9.6%, i.e. urban infrastructures account for approximately 34.5% of the territory sampled. Comparing this area to the 45.5% covered by lawns and the 15.7% covered by woods, the area occupied by urban infrastructures becomes even smaller.

The area covered by the urban vegetation canopy (woods and urban afforestation) amounts to 29.9% of the area studied (Figure 37). This canopy is made up of the area's urban forest and some fragments (Bosques) with denser vegetation, made up of small areas with denser tree planting and native vegetation (Bosque de vegetaçao and bosque de araucàrias), another with eucalyptus plantations (eucalyptus forest) and an even larger area, which is part of the fragmentary composition of the riparian forest of the small dam on the Monjolinho river, where pine plantations predominate with an understorey of regenerating native vegetation (pine forest). The forests within the studied area make up 52% of the canopy area. These fragments can maximize the benefits of afforestation, such as maintaining temperature, humidity and even conserving biodiversity, but more in-depth studies should be carried out in the area to understand the real importance and function of these urban fragments.

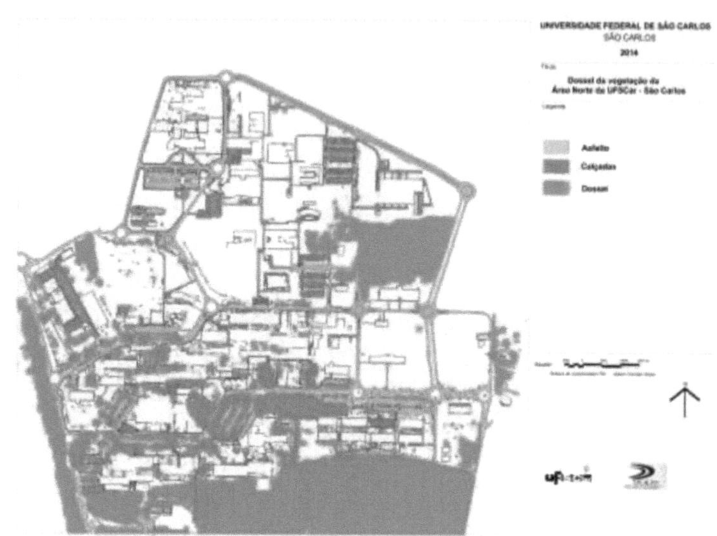

Figure 37: Canopy of urban vegetation, sidewalks and streets (asphalt) in the North Area of UFSCar - Sao Carlos. Source: Google Earth . Prepared by: Lessi, B. F. (2014).

The canopy found in the North Area of the *campus* also covers streets and sidewalks. The trees cover approximately 19% (21,078.02 m^2) of the asphalt area and 17% (13,580.31 m2) of the sidewalk area in the northern area of the *campus* (Figure 38). The low coverage of sidewalks and asphalt can be explained by (I) the fact that some of the trees are small and medium-sized, and therefore have smaller canopies (II) some of the trees are far from the streets and sidewalks. Several factors can affect the canopy cover of an urban area. Planning planting actions, preparing seedlings, phytosanitary control, pruning, etc., are important if the afforestation is to be integrated and perform its functions with quality.

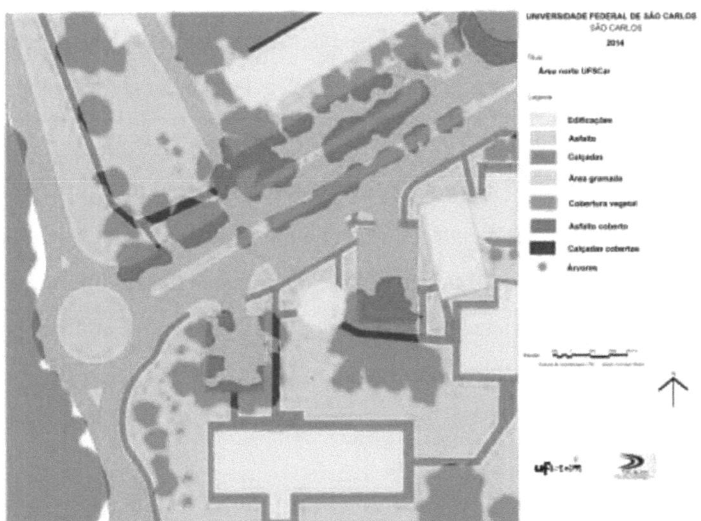

Figure 38: Coverage of sidewalks and streets (asphalt) by the canopy of urban vegetation in the North Area of UFSCar - Sao Carlos, in detail (Department of Mathematics). Source: EDF UFSCar (2013). Elaboration: Lessi, B. F. (2014).

Actions to plant trees and provide better coverage of streets and sidewalks by the canopy of urban vegetation should be considered. Thousands of people circulate *on campus* (MELÂO et al., 2011), walking along sidewalks and streets, so protection from UV rays and thermal comfort make walking more comfortable, safe and healthy.

According to Guzzo and team (2006) university *campuses* can be considered as potentially collective urban open space with the possibility of public use. It is also considered that urban open spaces when intended for environmental conservation and the implantation of vegetation are called green areas, public or private. According to the author, a green area should be made up of at least 70% vegetated areas with permeable soil. Another type of urban open space would be the urban park, which is larger than public squares and gardens and is intended for leisure, conserving natural resources and improving the city's environmental conditions.

So, in a way, the UFSCar - Sâo Carlos *campus*, with reference to the North Area (object of the study), is very similar to a potentially collective public green area, as it offers a vast vegetated area, reaching 61%, with 30% canopy cover. On the other hand, it can also resemble a large urban park, with leisure areas, conservation of natural resources and because it is in a peri-urban area and has a large vegetated area, contributing to the environmental quality of the city (COLDING, 2007).

The *campus* has even less urban area and infrastructure compared to an urban center, thus presenting fewer conflicts between urbanization and urban tree planting, as previously discussed in this paper. It is also much larger in area than public squares. In this way, its comparison is restricted to other campuses, which can also differ greatly in terms of area, urban infrastructure and vegetation.

CHAPTER 5

5 FINAL CONSIDERATIONS

The data collected shows that the urban forestation of the North Area of UFSCar - Sao Carlos has a high species richness. It also has a good distribution of species dominance, which is within the range recommended by Santamour (1990) to prevent the spread of pests. From this perspective, it can be said that the tree community in the North Area of UFSCar - São Carlos has considerable stability, showing advisable indicators of dominance, richness and diversity of advisable species.

The presence of native species, especially those of the cerrado, makes a major contribution to the preservation of this vegetation. The fact that 42% of the trees are zoochorous represents a great potential for attracting native fauna, especially birds. In addition, the existence of new plantations with a predominance of native species indicates compliance with the guidelines set out in the Institutional Development Plan (IDP).

Urban afforestation in the study area is predominantly made up of trees, which can maximize environmental benefits in the urban area, but they are mostly small and medium-sized. This factor could be improved, since larger trees can offer more benefits. The analysis of land use in the North Area of UFSCar - Sâo Carlos revealed large areas of lawn that can accommodate large species, which should be part of the afforestation to improve the environment.

The identification of the species present in the UFSCar North area indicated a large number of exotic species (50%), which should be avoided, especially with regard to *Ligustrum lucidum*, *Espathodea nilótica*, *Terminalia catappa* L., *Archontophoenix cunninghamiana* H. Wendl. &. Drude, Pinus spp, *Syzygium cumini* (L.) Skeels species with invasive potential. Although some local native species are found, taking into account that the university has a cerrado area attached to the *campus*, it is advisable to plant native trees of this vegetation physiognomy in its urban area, especially as they are adapted to the local climate and to possible attacks by pests and diseases.

The analysis of species distribution has proved decisive in the management and analysis of urban tree planting, since the simple assessment of richness and diversity can hide undue aggregations and susceptibility to pests and diseases. In this study, the four most abundant species have an aggregated distribution and should have their sites properly chosen, for the next plantings, away from the agglomeration. At the same time, because they have an aggregated distribution, they should receive special attention, since they represent 29.2% of the tree population and the appearance of a pest could represent a major loss.

In general, most of the individuals inventoried are in good condition and physical integrity. However, there is a lot of physical damage caused by pruning (53.7%), which can be a gateway for pests and diseases to settle in and affect a large proportion of the trees. The need for planning to determine phytosanitary control actions, risk assessment and identification of trees with low integrity is necessary, as is planning for future plantings, determining standards for seedling size, species, pits, etc.

The relationship between urban tree planting in the North Area of UFSCar - Sao Carlos and the *campus* infrastructure, even without a tree planting plan, is not very conflictive. However, these must be resolved and avoided in the future through planning. Decision-making guided by a plan can prevent management with drastic pruning that causes serious damage to the plants. In this context, the inventory revealed a lot of physical damage caused by pruning, which can contribute to the emergence of pests and diseases.

It is worth noting that, as it is a university, the urban area studied has a low density of buildings and electrical wiring, which may be a determining factor in the low number of conflicts. Another aspect is that most of the trees are in places without urban infrastructure, such as parking lots and grassy areas, which also reduces the incidence of conflicts even with sidewalks, where regardless of the species chosen and the behavior of their roots, they will not cause damage.

The analysis of the urban vegetation canopy revealed a coverage of 29.9%. It also indicated that this canopy is well divided between the sparse afforestation and the vegetation fragments (forests) within the area, with 48% and 52% of the canopy respectively, revealing the importance of the combination of these two forms of distribution for the environment. Sidewalks and streets had low percentages of area covered by the urban vegetation canopy (19% and 17% respectively) and should therefore be given greater attention in future plantings, in order to provide greater shading and thermal comfort for people.

UFSCar's PDI includes in its guidelines a concern for the environmental comfort of the space, guaranteeing the creation of urban parks with green areas and equipment for sport, leisure and culture, creating parks and gardens, maintaining $30m^2$ of green areas per inhabitant, always seeking to conserve native vegetation. With this in mind, UFSCar can be compared to a large urban park or a public green area.

The results of this research show that carrying out an inventory of urban forestry combined with a geographic information system (GIS) can generate information plans of great relevance to management. The database combined with the GIS makes it possible to manipulate, combine and localize the data, which allows for a spatial analysis of the data, accurately identifying the geographical location of each piece of information. The database formed in this work can be used as a diagnostic tool for drawing up a tree-planting plan, which in the medium and long term can help to improve environmental management.

The management of pruning waste should also be considered. The creation of a tree waste plan should be incorporated into the urban forestry plan. Stockpiling this material for drying can pose a risk of fire, especially if it is stored close to natural areas. One possible use would be to compost it into fertilizer, which would then be used to fertilize the urban trees that need it, as well as for seedling nurseries.

Drawing up an arborization plan, containing clear guidelines for activities and actions related to management, including making more coherent and informed decisions about arborization management, should address issues that go beyond those discussed in this work, raising the university as an example of planning and management of this important urban component.

BIBLIOGRAPHICAL REFERENCES

AGUIRRE JUNIOR, J. H.; LIMA, A. M. L. Use of trees and shrubs in Brazilian cities. **Revista da sociedade brasileira de arborizaçâo urbana**, v.2, n.4, dez. 2007, p. 50-66.

ALBERTI, M. **Advances in Urban Ecology: Integrating Humans and Ecological Processes in Urban Ecosystems**. Seattle, Washington, USA: University of Washington, Springer US, 2008. p. 379

ALBERTIN, R. et al. Quali-quantitative diagnosis of road tree planting in Nova Esperança-Paranà, Brazil. **Revista da sociedade brasileira de arborizaçâo urbana**, v. 6, n. 3, p. 128-148, 2011.

ALBRECHT, J. M. F. **Functional analysis, tree composition and management of the road network and green areas in the city of Sâo Carlos - SP**. Federal University of Sâo Carlos, UFSCar, Sâo Carlos, SP,. 1998. p. 217.

ALVEY, A. A. Promoting and preserving biodiversity in the urban forest. **Urban Forestry & Urban Greening**, v. 5, n. 4, p. 195-201, dec. 2006.

ANDREATTA, T. et al. Analysis of tree planting in the urban context of avenues in Santa Maria, RS. **Revista da sociedade brasileira de arborizaçâo urbana**, v. 6, n. 1, p. 36-50, 2011.

BANERJEE, B. et al. **ECOSYSTEM SERVICES PLAN: Yale University School of Medicine *Campus***. Yale, USA: Yale School of Forestry and Environmental Studies, 2011. p. 37

BENATTI, D. et al. ARBOR-URBAN INVENTORY OF THE MUNICIPALITY OF SALTO DE PIRAPORA, SP. **Revista Arvore**, v. 36, n. 5, p. 887-894, 2012.

BIAS, E. D. S.; BAPTISTA, G. M. D. M.; LOMBARDO, M. A. Analysis of the urban heat island phenomenon using a combination of landsat and ikonos data. **Anais XI SBSR, Belo Horizonte, Brazil**, p. 1741-1748, 2003.

BOLUND, P.; HUNHAMMAR, S. Ecosystem services in urban areas. **Ecological**

Economics, v. 29, n. 2, p. 293-301, May 1999.

BORTOLETO, S. **Quali-quantitative inventory of roadside trees in the town of Aguas de Sâo Pedro-SP**. Piracicaba, SP: Escola Superior de Agricultura "Luiz de Queiroz", Universidade de Sao Paulo - USP, 2004. p. 99

BRITT, C.; JOHNSTON, M. **Trees in Towns II A new survey of urban trees in England and their condition and management**. London: Communities and Local Government, 2008. p. 36

BUENO, E. S.; XIMENES, D. S. S. The importance of green infrastructure in environmental design: a study of the university city area and the Butanta Institute. **Revista LABVERDE**, n. 3, p. 128-154, 2011.

CADORIN, D.; SILVA, L.; HASSE, I. Tree-planting characteristics of the Cadorin, Parzianello and La Salle neighborhoods in Pato Branco-PR (2007). **Revista da sociedade brasileira de arborizaçâo urbana**, v. 4, n. 2007, p. 40-52, 2008.

CARNEIRO, D. P. Q. et al. Heat islands on the UNICAMP *campus*. **Revista Ciências do Ambiente On-Line**, v. 3, n. 2, p. 43-48, 2007.

CEMIG. **Afforestation Manual**. Belo Horizonte: Companhia Energética de Minas Gerais, Fundaçâo Biodiversitas, 2011. v. 28p. 112

COLDING, J. "Ecological land-use complementation" for building resilience in urban ecosystems. **Landscape and Urban Planning**, v. 81, n. 1-2, p. 46-55, May 2007.

CONDEMA. Municipal Council for the Defense of the Environment. **COMDEMAS/SC Resolution No. 01/2012**. Sâo Carlos City Hall, 2012

COUTO, C. D. S. **Inventory and diagnosis of urban afforestation in the Benfica neighborhood, municipality of Rio de Janeiro, RJ.** Seropédica, RJ: Federal Rural University of Rio de Janeiro, 2006. p. 41

DUARTE, F. G. et al. Termites (Insectta: Isoptera) in urban trees in Zone 1 of Maringà - PR. **Revista em Agronegócios e Meio Ambiente**, v. 1, n. 1, p. 8799, 2008.

ELMQVIST, T. et al. **Urbanization, Biodiversity and Ecosystem Services: Challenges and Opportunities: A Global Assessment.** Springer, 755p. 2013

FALCE, B. DE O. et al. Analysis of the spatial distribution of trees and shrubs in terms of size, taxonomy and use using the Geographic Information System. **Revista da sociedade brasileira de arborizaçâo urbana**, v. v, n. 1, p. 23-34, 2012.

FALEIRO, W. Arborization of the Umuarama *campus of* the Federal University of Uberlândia. **Revista Cientifica Eletrônica de Engenharia Florestal**, v. 10, 2007.

FARIA, J.; MONTEIRO, E. A.; FISCH, S. T. V. Arborization of public roads in the municipality of Jacarei-SP. **Revista da sociedade brasileira de arborizaçâo urbana**, v. 2, n. 4, p. 20-33, 2007.

FRISCH, J. D.; FRISCH, C. D. Brazilian birds and the plants that attract them. Sao Paulo: Ed. Dalgas Ecoltec - Ecologia Técnica Ltda. 2005, p. 480.

GILL, S. E. et al. Adapting Cities for Climate Change : The Role of the Green Infrastructure. **Bult Envirinment**, v. 33, n. 1, p. 115-133, 2007.

GUZZO, P.; CARNEIRO, R. M. A. **Vamos Arborizar Ribeirâo Preto**. 1ª ed. Prefeitura Municipal de Ribeirao Preto, Secretaria Municipal do Meio Ambiente, Ribeirao Preto, SP, 2008. p. 40.

GUZZO, P.; CARNEIRO, R. M. A.; OLIVEIRA JÙNIOR, H. DE. Municipal register of urban open spaces in Ribeirao Preto (SP): Public access, indices and basis for new management tools and mechanisms. **Revista da sociedade brasileira de arborizaçâo urbana**, v. 1, n. 1, p. 19-30, 2006.

Global Invasive Species Database (http://www.issg.org/database), accessed on 05/02/2015.

GÓMEZ-BAGGETHUN, E. et al. Urban Ecosystem Services. In: ELMQVIST, T. et al. (Eds.). **Urbanization, Biodiversity and Ecosystem Services: Challenges and Opportunities**. Dordrecht: Springer Netherlands, 2013.

GRACIOLI, C. R. et al. Afforestation of the *campus of* the Federal University of Santa

Maria and awareness of the academic community. **Revista Monografias Ambientais**, v. 3, n. 3, p. 421-429, 2011.

GÜNTHER, H. Proposal for the Ecological Recovery of a University *Campus*: Green UnB Program. **Textos de Psicologia Ambiental**, n. 25, p. 1-3, 1994.

HAMMER, 0.; HARPER, D. A. T.; RYAN, P. D. PAST: Paleontological statistics sofware package for education and data analysis. **Palaeontologia Electronica**, v. 4, n. 1, p. 1-9, 2001.

HIEMSTRA, J. A.; BIJL, E. S.- VAN DER; TONNEIJCK, A. E. G. **Trees: relief for the city**. Netherlands: Plant Publicity Holland, 2008.

IBGE, at http://cidades.ibge.gov.br/painel/painel.php?lang=&codmun=354890&search=s ao-paulo|sao-carlos|infograficos:-dados-gerais-do-municipio, accessed on 04/11/14.

IBGE, Demographic Census 2010, in

http://cidades.ibge.gov.br/xtras/perfil.php?lang=&codmun=354890&search=||inf ogr%E 1 ficos:-informa%E7%F5es-completas, accessed on 04/11/14.

HORUS INSTITUTE. National database of invasive alien species, I3N

Brazil, Instituto Hórus de Desenvolvimento e Conservaçao Ambiental, Florianópolis - SC. http://i3n.institutohorus.org.br, accessed on 05/11/2014).

KONIJNENDIJK, C. C. et al. **Urban Forests and Trees**. Netherlands: Springer

Verlag Berlin Heidelberg, 2005. p. 525

KONIJNENDIJK, C. C. et al. Defining urban forestry - a comparative perspective of North America and Europe. **Urban Forestry & Urban Greening**, v. 4, n. 3-4, p. 93-103, 2006.

KUHLMANN, M. **Frutos e sementes do Cerrado atractativos para fauna: guia de campo**. Brasilia: Ed. Rede de Sementes do Cerrado, 2012, p. 360.

KURIHARA, D.; IMANA-ENCINAS, J.; PAULA, J. Survey of the arborization of the *campus of* the university of Brasilia. **Cerne**, v. 11, n. 2, p. 127-136, 2005.

KURIHARA, D. L.; ENCINAS, J. I. Analysis of the afforestation of the *Campus of* the University of Brasilia using ikonos images. **Brasil Florestal**, v. 78, p. 81-87, 2003.

LEAL, L.; PEDROSA-MACEDO, J.; BIONDI, D. Census of the arborization *of Campus* III-Polytechnic Center of the Federal University of Paranà. **Scientia Agraria**, v. 10, n. 6, p. 443-453, 2009.

LIMA NETO, E. M. DE; BIONDI, D.; ARAKI, H. Aplicação do SIG na arborização viaria - unidade amostral em Curitiba-PR. **III Brazilian Symposium on Geodetic Sciences and Geoinformation Technologies**, p. 1-6, 2010.

LOMBARDI, J. A.; MORAIS, P. O. Levantamento floristico das plantas empregadas na arborizaçâo da *campus* da Universidade Federal de Minas Gerais , Belo. v. 4, n. 2, p. 83-88, 2003.

LORENZI, H. **Arvores brasileiras: manual de identificaçâo e cultivo de plantas arbóreas do Brasil**, vol. 1, 5ª ed., p. 384, Nova Odessa, SP: Instituto Plantarum, 2008.

LORENZI, H. **Arvores brasileiras: manual de identificaçâo e cultivo de plantas arbóreas do Brasil**, vol. 2, 3rd ed., p. 384, Nova Odessa, SP: Instituto Plantarum, 2009a.

LORENZI, H. **Arvores brasileiras: manual de identificaçâo e cultivo de plantas arbóreas do Brasil**, vol. 3, 1st ed., p. 384, Nova Odessa, SP: Instituto Plantarum, 2009b.

LORENZI, H. et al. **Arvores exóticas no Brasil: Madeireiras, ornamentais e aromâticas**, F ed. p. 361, Nova Odessa, SP: Instituto Plantarum, 2003.

MARTINS, L. et al. RELATIONSHIP BETWEEN PODAS AND PHYTOSANITARY ASPECTS IN URBAN TREES IN THE CITY OF LUIZIANA, PARANA. **Revista da sociedade brasileira de arborizaçâo urbana**, v. 5, n. 4, p. 141-155, 2010.

MARTINS, L. F. V.; ANDRADE, H. H. B. DE; ANGELIS, B. L. D. DE. Relationship between pruning and physical and health aspects of urban trees in the city of Luiziana, Paranâ. **Revista da sociedade brasileira de arborizaçâo urbana**, v. 5, n. 4, p. 141-

155, 2010.

MASCARÓ, L.; MASCARÓ, J. L. Vegetaçao Urbana. 3rd ed., Porto Alegre, RS: Masquatro Editora, 2010. p. 212.

MATOS, E.; QUEIROZ, L. P. DE. **Trees for Cities**. 1 ed. Salvador, BA: Solisluna, 2009. p. 340.

MATTHEWS, S. **GISP**. 1 ed. ed. South America: Global Invasive Species Program, 2005. p. 80.

MAZA, C. L. DE LA et al. Vegetation diversity in the Santiago de Chile urban ecosystem. **Arboricultural Journal**, v. 26, n. 4, p. 347-357, dec. 2002.

MAZIOLI, B. Inventory and diagnosis of urban afforestation in two neighborhoods in the city of Cachoeiro do Itapemirim, ES. **Monograph**, p. 53, 2012.

MCPHERSON, E. GREGORY; NOWAK, DAVID J.; ROWNTREE, ROWAN A. Chicago's urban forest ecosystem: results of the Chicago Urban Forest Climate Project, **General Technical Report NE-186,** 201 p. 1994.

MELÂO, M. DA G. G. et al. Diagnosis and environmental characterization UFSCar, Sao Carlos *campus*. **Report presented to the Federal Public Prosecutor's Office and CETESB**, p. 61, Sept. 2011.

MELO, E. F. R. Q.; SEVERO, B. M. A. Tree vegetation on the *campus* of the University of Passo Fundo. **Revista da sociedade brasileira de arborizaçao urbana**, v. 2, n. 2, p. 76-87, 2007.

MENEGHETTI, G. I. P. **Study of two sampling methods for an inventory of street trees in the seafront neighborhoods of the municipality of Santos, SP**. Piracicaba: Escola Superior de Agricultura "Luiz de Queiroz", Universidade de Sao Paulo - USP, 2003. p. 114.

MILANO, M. S. **Quali-quantitative evaluation and management of urban tree planting: an example from Maringà - PR**. Curitiba, PR: Universidade Federal do Paranà, 1988. p. 136

MIRANDA, T. O. DE; CARVALHO, S. M. Quantitative and qualitative survey of tree individuals present in the streets of the Ronda neighborhood in Ponta Grossa - PR. **Revista da sociedade brasileira de arborizaçao urbana**, v. 4, n. 3, p. 143157, 2009.

MOTTA-JUNIOR, J. C. et al. Diet of the maned wolf, Chrysocyon brachyurus , in central Brazil. **Journal of Zoology**, v. 240, n. 2, p. 277-284, Oct. 1996.

MOTTA-JUNIOR, J. C.; VASCONCELOS, L. A. DA S. Survey of the birds of the *campus of* the Federal University of Sao Carlos, State of Sao Paulo, Brazil. **Anais do VII Seminàrio Regional de Ecologia**, v. 7, p. 159-171, 1996.

NA, H. R. et al. Modeling of urban trees' effects on reducing human exposure to UV radiation in Seoul, Korea. **Urban Forestry & Urban Greening**, Jun. 2014.

NICHOLSON-LORD, D. **The Greening of the Cities**. e-Book edi ed. London, UK: Taylor & Francis e-Library, 2005. p. 288

NOWAK, D. J. et al. Measuring and analyzing urban tree cover. **Landscape and Urban Planning**, v. 36, n. 1, p. 49-57, oct. 1996.

NOWAK, D. J. et al. A ground-based method of assessing urban forest structure and ecosystem services. **Arboriculture & Urban Forestry**, v. 34, n. November, p. 347-358, 2008.

NOWAK, D. J.; CRANE, D. E. Carbon storage and sequestration by urban trees in the USA. **Environmental Pollution**, v. 116, p. 381-389, 2002.

NOWAK, D. J.; CRANE, D. E.; STEVENS, J. C. Air pollution removal by urban trees and shrubs in the United States. **Urban Forestry & Urban Greening**, v. 4, n. 3-4, p. 115-123, abr. 2006.

NUFU. **Trees Matter ! Bringing lasting benefits to people in towns**. London: National Urban Forestry Unite, 2005. p. 20.

OLIVEIRA FILHO, P. C. DE; KÜSTER DA SILVA, S. V. An information system for spatial and decision-making support for urban tree management in the municipality of Guarapuava, Paranà. **Revista da sociedade brasileira de arborizaçao urbana**, v. 5,

n. 3, p. 82-96, 2010.

OLUDUNFE, S. O. **Urban Forest Management Plan**. San Diego, California, USA: Trees *Campus* USA, 2011. p. 64.

PAIVA, A. DE. Aspects of urban afforestation in the center of Cosmópolis-SP. **Revista da sociedade brasileira de arborizaçao urbana**, v. 4, n. 4, p. 17-31, 2009.

PIRES, N. et al. Urban afforestation in the municipality of Goiandira/GO - qualitative and quantitative characterization and management proposal. **Revista da sociedade brasileira de arborizaçao urbana**, v. 5, n. 3, p. 185-205, 2010.

PME. **Urban Forestry Master Plan for the Municipality of Erechim, RS**. 1ª Edition ed. Erechim - RS: Prefeitura Municipal de Erechim, 2011. p. 120.

PMSC. **Sao Carlos Urban Afforestation Master Plan (PDAU)**. Sao Carlos, SP: Decree 216, Sao Carlos City Hall, 2009. p. 10.

PRB. **WORLD POPULATION DATA SHEET**. Washington, USA: Population Reference Bureau, 2014. p. 20.

PSP. **Technical Manual on Tree Pruning**. Sao Paulo, SP: Prefeitura da Cidade de Sao Paulo, Secretaria do Verde e do Meio Ambiente, 2002. p. 32.

PSP. **Manual Técnico de Arborizaçao Urbana**. 2a Ediçao ed. Sao Paulo: Prefeitura da Cidade de Sao Paulo, Secretaria do Verde e do Meio Ambiente, 2005. p. 48.

ROCHA, R. DA; LELES, P.; NETO, S. Arborization of public roads in Nova Iguaçu, RJ: the case of the Rancho Novo and Centro neighborhoods. **Revista Àrvore**, v. 28, n. 4, p. 599-607, 2004.

ROSSATTO, D.; TSUBOY, M. S. F.; FREI, F. Urban afforestation in the city of Assis-SP: a quantitative approach. **Revista da sociedade brasileira de arborizaçâo urbana**, v. 1, n. 3, p. 1-16, 2008.

SANTAMOUR, F. Trees for urban planting: diversity uniformity, and common sense. **7th Conference of the Metropolitan Tree Improvement Alliance**, v. 7, p. 57-66, 1990.

SCHUCH, M. I. S. **Urban afforestation: a contribution to quality of life using geotechnologies**. Santa Maria, RS: Federal University of Santa Maria, 2006. p. 102.

SILVA, E. M. DA et al. Study of urban afforestation in the Mansour neighborhood, in the city of Uberlândia - MG. **Caminhos de Geografia**, v. 3, n. 5, p. 73-83, 2002.

SILVA FILHO, D. F.; BORTOLETO, S. Use of diversity indicators in the definition of a road tree management plan for Aguas de Sâo Pedro - SP. **Revista Arvore**, v.29, n.6, p.973-982, 2005.

SILVA FILHO, D. F. DA; COSTA, F. P. DA S.; POLIZEL, J. Planning urban afforestation in the city of Engenheiro Coelho-SP: Use of GIS and sample inventory. **Revista Geogràfica em Atos**, v. 1, n. 12, p. 1-8, 2012.

SILVA, L. **Situação da arborização viària e proposta de espécies para os bairros Antônio Zanaga I e II, da cidade de Americana/SP**. Piracicaba: Luiz de Queiroz College of Agriculture, 2005. p. 81.

SPADOTTO, L. G. F.; DELMANTO JÛNIOR, O. Planning and georeferencing urban afforestation using geoprocessing techniques. **Tékhne ε Lógos**, v. 1, n. 1, 2009.

STRANGHETTI, V.; SILVA, Z. A. V. DA. Diagnosis of the afforestation of public roads in the municipality of Uchôa - SP. **Revista da sociedade brasileira de arborizaçâo urbana**, v. 5, n. 2, p. 124-138, 2010.

SUCOMINE, N. **Characterization and analysis of the arboreal heritage of the central urban road network of the municipality of Sâo Carlos-SP**. Sâo Carlos, SP: Federal University of Sâo Carlos, 2010. p. 111.

TOLEDO, M. C. B. DE. **Analysis of urban green areas at different scales for bird conservation**. Botucatu, SP: Universidade Estadual Paulista "Jùlio de Mesquita Filho" - UNESP, 2006. p. 1-149.

TOSCAN, M. et al. Inventory and analysis of the afforestation of the Vila Yolanda neighborhood, in the municipality of Foz do Iguaçù-PR. **Revista da sociedade brasileira de arborizaçao urbana**, v. 5, n. 3, p. 165-184, 2010.

UFSCAR. **Institutional Development Plan - PDI**. Sao Carlos, SP: Federal University of Sao Carlos, 2013. p. 50.

URBAN, J. R. Evaluation of Tree Planting Practices in the Urban Landscape. In: **Make Our Cities Safe for Trees: Proc. 4th Urban Forestry Conference**. Washington, DC: The American Forestry Association, 1989. p. 119-127.

Printed by Books on Demand GmbH, Norderstedt / Germany